Handbook of Molecular Biology and Biochemistry

Handbook of Molecular Biology and Biochemistry

Edited by **Samantha Granger**

SYRAWOOD
PUBLISHING HOUSE

New York

Published by Syrawood Publishing House,
750 Third Avenue, 9th Floor,
New York, NY 10017, USA
www.syrawoodpublishinghouse.com

Handbook of Molecular Biology and Biochemistry
Edited by Samantha Granger

International Standard Book Number: 978-1-68286-011-3 (Hardback)

Contents

Preface

Molecular biology and biochemistry are two complementary disciplines which help us to understand diverse biological processes. This book aims to elaborate the current developments in cell biology, microbiology, bioengineering, etc. It aims to broaden the horizon of research with the help of advanced inputs by eminent experts. This book elucidates the concepts and innovative models around prospective developments in these fields and will be beneficial for students and academicians pursuing biochemistry, molecular biology, and allied sciences.

All of the data presented henceforth, was collaborated in the wake of recent advancements in the field. The aim of this book is to present the diversified developments from across the globe in a comprehensible manner. The opinions expressed in each chapter belong solely to the contributing authors. Their interpretations of the topics are the integral part of this book, which I have carefully compiled for a better understanding of the readers.

At the end, I would like to thank all those who dedicated their time and efforts for the successful completion of this book. I also wish to convey my gratitude towards my friends and family who supported me at every step.

Editor

Haematological Response and Serum Biochemical Profile of Broiler Finishers Fed with Oxytetracycline and Stonebreaker (*Phyllanthus amarus*) Leaf Meal

A. D. Ologhobo[1] and I. O. Adejumo[2*]

[1]*Department of Animal Science, University of Ibadan, Ibadan, Nigeria.*
[2]*Department of Animal Science, Landmark University, Omu-Aran, P.M.B.1001, Kwara State, Nigeria.*

Authors' contributions

This work was carried out in collaboration between both authors. Author ADO designed the study and carried out the feeding trial. Author IOA managed the literature search, statistical analysis and wrote the first manuscript. Both authors approved the final manuscript.

Editor(s):
(1) Laura Pastorino, Depart. Informatics, Bioengineering, Robotics and Systems Engineering (DIBRIS), University of Genoa, Italy.
Reviewers:
(1) Anonymous, Ghana.
(2) Anonymous, Brazil.
(3) Anonymous, Brazil.
(4) Anonymous, Nigeria.
(5) Anonymous, Nigeria.

ABSTRACT

Aims: The aim of the study was to assess the effect of *Phyllanthus amarus* leaf meal on haematology and serum biochemical profile of broiler finishers vis a vis oxytetracycline, in order to serve a basis for further study focusing on antibacterial properties of *Phyllanthus amarus*.
Study Design: The design of the study was completely randomised design.
Place and Duration of Study: The study was carried out at the Teaching and Research Farm, University of Ibadan, Nigeria. The study lasted for 6 weeks.
Methodology: One hundred and eight mixed-sex (Hybro PG) four-week old chicks were used for the study.
Four dietary treatments were formulated. Each treatment had three replicates, while each replicate had nine birds. The experimental diets contained 0.25% of tetracycline (T1), 0.20%, 0.40%, and 0.60% of *Phyllanthus amarus* leaf meal for T2, T3, and T4 respectively.
Results: Except for packed cell volume (PCV) and haemoglobin (Hb), there were no significant

Corresponding author: Email: smogisaac@gmail.com

differences across the treatments for all the haematological parameters measured. There were no significant differences across the treatments for all the serum biochemical profile measured except for albumin and alanine amino transferase (ALT). There was significant ($P=.05$) increase for albumin for all diets containing Phyllanthus amarus leaf meal. The highest mean value was recorded for birds in T3 (1.56 g/dl), followed by those in T2 (1.45 g/dl). Those in control diet (T1) had the least mean value (0.53 g/dl). ALT did not follow a specific pattern.
Conclusion: Phyllanthus amarus leaf meal in the diets of broiler finisher chickens did not pose any health hazards to the haematological parameters and serum biochemical profile of the broiler finishers at the levels of inclusion.

Keywords: Antibiotic; blood; broilers; Phyllanthus amarus.

1. INTRODUCTION

Poultry meat has been regarded as the second largest global food commodity [1]. High cost of feed ingredients and disease outbreak are major factors stressing poultry production in the tropics. Hence, the need for use of cheap feed ingredients and antibiotics. Antibiotics have been defined as natural metabolites of fungi, bacteria or yeast that inhibit bacteria proliferation by direct action, bactericidal effect or bacteriostatic effect on sensitive fungi and/or bacteria [2]. However, synthetic antibiotic may leave resistance and is also becoming unaffordable for livestock farmers. This then necessitates the need to look inward for alternative antibiotics in leaves and other plant parts that could be included as additives. Finding healing power in plants is an age long idea. The use of leaf meal of monogastric animals is however limited, due to inability of the animals to utilise fibrous materials like ruminants. In a study conducted by Ogbuewu et al. [3], neem leaf meal was reported to have mild depressive effect on haemoglobin (Hb) concentration and packed cell volume (PCV) of female rabbits. Esonu et al. [4] on the other hand reported slight increments in the values of Hb and PCV of laying hens fed neem leaf meal diets. Microdesmis puberula leaf meal had been reported [5] to significantly improve feed intake, body weight gain, feed conversion ratio, nutrient utilization and organ characteristics of broilers at 0%, 10%, and 15% inclusion levels. The result of the study carried out by James et al. [6] showed that aqueous extract of Phyllanthus amarus significantly affected PCV and Hb concentrations of albino rats. The variations were however within the normal range.

Phyllanthus sp. is used among indigenous peoples as a small annual herb to cure many diseases such as hepatitis and other liver diseases [7]. Previous studies had indicated that Phyllanthus sp has anticancer effect as well as the ability to reverse the negative impacts of

aflatoxicosis of broiler chickens [8,9]. It has also been reported that Phyllanthus amarus could be used as a medicinal plant to treat hepatic disease, eye infection, snakebites and acnes [10]. Phyllanthus amarus extract has been suggested as having the potential to reduce the effects of fumonisins on liver [11]. Phyllanthus amarus, like other tropical tree leaves contains some bioactive compounds which may affect nutrient utilization, haematological and serum biochemical parameters in animals [3,12,13]. The anti-nutritional compounds contained in Phyllanthus amarus include oxalate, phytate, hydrogen cyanide, nitrate and tannin [14]. There is need to look for alternative antibiotic for inclusion in poultry production, but the effect of the proposed alternatives need to be tested on haematology and serum biochemical before testing their antibiotic effect. Hence, this study was conducted to assess the effect of Phyllanthus amarus on blood parameters of broiler finishers.

2. MATERIALS AND METHODS

2.1 Location and Duration of the Study

The study was carried out at the Poultry Unit of Teaching and Research Farm, University of Ibadan, Nigeria. The study lasted for 28 days.

2.2 Experimental Diets and Animals

One hundred and eight mixed-sex (Hybro PG) four-week old chicks were used for the study and were fed for 28 days with test diets. Four dietary treatments were formulated. Each treatment had three replicates, with nine birds per replicate. The design of the study was a completely randomised design. The birds were raised in a deep litter system, under environmental temperatures (23°C). Phyllanthus amarus used for the study was obtained from University of Ibadan, Nigeria. Ibadan ranges in elevation from 150 m in the valley area, to 275 m above sea level. It has a tropical wet and dry climate. It has a lengthy wet season with relatively constant temperatures throughout the year. It has the mean total rainfall of about 1420.06 mm, falling in approximately 109 days. There are two peaks for rainfall, June and September. The mean maximum temperature is 26.46°C, minimum 21.42°C and the relative humidity is 74.55%. The harvested plants were air dried under shade until they were crispy to touch, while still retaining their greenish colour. The leaves were not sun dried to prevent destruction of bioactive molecules of the plant at high temperature. The dried leaves were milled using a kitchen blender to produce stonebreaker (Phyllanthus amarus) leaf meal (SBLM), suitable for incorporation into broiler diets. The diets contained 0.25% of tetracycline (T1), 0.20% of SBLM (T2), 0.40% of SBLM (T3), and 0.60% of SBLM (T4). Feed intake was determined by deducting the weight of left over from the weight of feed fed.

2.3 Data Collection and Analysis

At the end of the study, two birds were randomly selected from each replicate and blood samples were collected from them. Blood samples were collected from the birds using new, sterilized syringe and needles through their jugular veins.

Blood samples were collected before the birds were served feed for the day. Blood samples for haematological analysis were collected into sterilized glass tubes containing ethylene diamine tetra acetic acid (EDTA) as anticoagulant. Samples for serum biochemical analysis were collected into tubes without EDTA. Packed cell volume (PCV) was determined using micro haemotocrit and haemoglobin (Hb) concentrations were determined using cyanmethaemoglobin method [15]. Erythrocyte (RBC) and leukocytes (WBC) counts were determined using the improved neubrauer haemocytometer after appropriate dilution. Blood samples for serum biochemical analysis after being centrifuged were taken to the laboratory for the analysis. Aspartate amino transferase (AST) and alanine amino transferase (ALT) were determined using spectrophotometric methods. Proximate analysis was carried out using the standard procedures of Association of Official Analytical Chemists [16]. Gross composition of the experimental diets is shown in Table 1.

2.4 Statistical Analysis

Data obtained were analysed using one way analysis of variance, and significant means were separated using Duncan multiple range test [17].

Table 1. Gross composition of experimental diets (g/100% DM)

Ingredients	T1	T2	T3	T4
Maize (%)	65.50	65.80	65.60	65.40
Soya bean meal (%)	20.00	20.00	20.00	20.00
Groundnut cake (%)	8.25	8.00	8.00	8.00
Fish meal (%)	2.00	2.00	2.00	2.00
Dicalcium phosphate (%)	1.50	1.50	1.50	1.50
Limestone(%)	1.50	1.50	1.50	1.50
Salt (%)	0.25	0.25	0.25	0.25
EoD (%)	0.25	0.25	0.25	0.25
L-Lysine (%)	0.25	0.25	0.25	0.25
DI-Methionine (%)	0.25	0.25	0.25	0.25
Oxytetracycline (%)	0.25	-	-	-
Phyllanthus amarus leaf meal (%)	-	0.20	0.40	0.60
Calculated nutrients				
Metabolisable energy (Kcal/kg	3064.00	3074.00	3072.00	3071.00
Proximate composition				
Dry matter	86.90	86.30	86.76	87.45
Crude protein (%)	19.69	20.13	20.13	21.88
Ether extract (%)	8.90	8.90	8.90	8.90
Ash (%)	12.00	13.00	13.00	13.00
Crude fiber (%)	4.20	3.70	3.70	3.70

EOD = Enrichment of diet contained vitamin A (10000000 UI), vitamin D3 (2000 IU), vitamin E (40000 mg), vitamin K (2000 mg), Vitamin B1 (1500 mg), vitamin B2 (4000 mg), Niacin (40000 mg), panthothenic (10000 mg), folic (1000 mg) (400000 mg), vitamin B6 (100 mg), vitamin B12 (20 mg), biotin (100 mg), choline chloride (300000 mg), manganese (80000 mg), iron (40000 mg), zinc (60000 mg), copper (80000 mg), iodine (30 mg), cobalt (300 mg), selenium (200 mg), antioxidant (100000 mg)

3. RESULTS AND DISCUSSION

Tables 2 and 3 show the haematological parameters and serum biochemical profile of broiler finishers fed experimental diets. The mean values for all the haematological parameters fell within the reported normal range. Except for packed cell volume (PCV) and haemoglobin (Hb), there were no significant differences across the treatments for all the haematological parameters measured. Birds on T4 recorded the least mean value for PCV (26.6%), while those on T2 recorded the highest mean value (36.33%). Hb followed the same pattern. Birds fed diets containing *Phyllanthus amarus* leaf meal recorded higher mean values for red blood cell counts and white blood cell counts, while lower values were recorded for lymphocytes.

There were no significant differences across the treatments for all the serum biochemical profile measured except for albumin and alanine amino transferase (ALT). There was significant (P=.05) increase for albumin for all diets containing *Phyllanthus amarus* leaf meal. The highest mean value was recorded for birds in T3 (1.56 g/dl), followed by those in T2 (1.45 g/dl). Those in control diet (T1) had the least mean value (0.53 g/dl). ALT did not follow a specific pattern. Birds in T4 had the least mean value (4.83 IU/l) which was statistically similar to those on T1 (6.98 IU/l) and T3 (5.23 IU/l), but statistically (P=.05) lower than those in T2 (8.48 IU/l). Birds on diets containing *Phyllanthus amarus* leaf meal recorded lower mean values for low density lipoprotein (LDL) and triglycerides. Mean values for cholesterol did not follow a specific pattern, but birds on T2 recorded the least mean value (122.80 mg/l), while those on T3 (157.89 mg/l) had the highest mean value. Mean values for total protein, bilirubin, AST, urea and high density

lipoprotein (HDL) were higher for birds fed diets containing *Phyllanthus amarus* leaf meal.

Blood parameters have been observed to be important factors used in assessing the response of animals to the diets they are fed [18,19]. The result of the study previously carried out by Ogbuewu [3] indicated that neem leaf meal had mild depressive effect on haemoglobin (Hb) concentration and packed cell volume (PCV) of female rabbits. James et al. [6] showed that aqueous extract of *Phyllanthus amarus* significantly affected PCV and Hb concentrations of albino rats. The result of this study indicated that all the haematological parameters measured fell within the normal (reference range) for healthy chickens [15], which suggested that the diets were well tolerated by the experimental animals. The increase in AST noticed in diets containing *Phyllanthus amarus* leaf meal is similar to the result of the study by Adedapo et al. [20] who reported significant increase in the levels of ALT and AST for rats fed diets containing 400 mg/kg and 1600 mg/kg doses of extracts of *Moringa oleifera* leaves. However, ALT in this study did not follow a similar trend with the result of Adedapo et al. [20]. ALT is known to increase in liver disease [20,21]. However, the increase in T2 in this study was not significantly different from the control diet. So, no liver damage is envisaged. Damaged liver would have been responsible for leakages of the serum enzymes into the blood. However, liver damage could not be suggested in this study as indicated in the result of the study. The result of this study therefore showed that inclusion of *Phyllantus amarus* at the levels of inclusion did not pose any adverse effects on the haematological parameters and serum biochemical profile of the experimental animals as there was no indication of organ toxicity from the serum enzyme assessed.

Table 2. Haematological parameters of broiler finisher chickens fed graded levels of *Phyllanthus amarus* leaf meal

Parameters	T1	T2	T3	T4	Reference values*
PCV (%)	31.00(±0.12)[b]	36.33(±1.49)[a]	31.67(0.08)[b]	26.60(±1.45)[c]	29.00- 44.00
Hb (g/l)	10.33(±0.04)[b]	12.11(±0.49)[a]	10.53(±0.02)[b]	8.89(±0.48)[c]	9.10- 13.90
RBC (X 10^6/l)	3.16(±0.17)	3.93(±0.07)	4.01(±0.09)	3.74(±0.01)	1.58 – 4.10
WBC (X 10^3/l)	16.42(±0.28)	17.00(±0.10)	18.67(±0.40)	17.25(±0.03)	9.20- 31.00
Monocytes (X10^6/µl)	3.00(±0.10)	2.00(±0.20)	2.67(±0.001)	3.00(±0.10)	0.06- 5.00
Lymphocytes (X10^6/µl)	70.67(±0.83)	69.00(±0.33)	65.00 (±0.88)	67.00(±0.28)	47.20- 85.00

*Means with different superscripts on the same rows indicate significant difference (P=0.05); T1= contains 0.25% oxytetracycline; T2= contains 0.20% of Phyllanthus amarus leaf meal; T3= contains 0.40% of Phyllanthus amarus leaf meal; T4= 0.60% of Phyllanthus amarus leaf meal; * Mitrika and Rawnsley (1977)*

Table 3. Serum biochemical profile of broiler finisher chickens fed with graded levels of *Phyllanthus amarus* leaf meal

Parameters	T1	T2	T3	T4
Total protein (g/dl)	1.89(±0.32)	2.83(±0.06)	3.92(±0.24)	3.57(±0.14)
Albumin (g/dl)	0.53(±0.20)^b	1.45(±0.08)^a	1.56(±0.11)^a	1.27(±0.02)^a
Urea (mg/dl)	3.77(±0.14)	4.18(±0.01)	4.60(±0.12)	4.39(±0.06)
Creatinine (mg/dl)	0.66(±0.01)	0.62(±0.01)	0.58(±0.02)	0.68(±0.01)
AST (IU/l)	52.26(±2.35)	54.42(±1.70)	68.34(±2.50)	65.18(±1.55)
ALT (IU/l)	6.98(±0.18)^{ab}	8.48(±0.63)^a	5.23(±0.35)^{ab}	4.83(±0.41)^b
Cholesterol (mg/l)	148.10(±0.82)	122.80(±6.81)	157.89(±3.77)	152.72(±2.21)
Triglyceride (mg/l)	171.70(±5.04)	152.10(±0.87)	151.89(±0.93)	144.25(±3.24)
Bilirubin (mg/dl)	0.12(±0.06)	0.18(±0.01)	0.17(±0.003)	0.18(±0.01)
HDL (mg/l)	12.89(±3.30)	21.53(±0.69)	31.39(±2.28)	29.48(±1.71)
LDL (mg/l)	34.34(±1.05)	30.42(±0.14)	30.31(±0.17)	28.40(±0.75)

Means with different superscripts on the same rows indicate significant difference (P=0.05); T1= contains 0.25% oxytetracycline; T2= contains 0.20% of Phyllanthus amarus leaf meal; T3= contains 0.40% of Phyllanthus amarus leaf meal; T4= 0.60% of Phyllanthus amarus leaf meal; HDL=high density lipoprotein; LDL= Low density lipoprotein; AST= aspartate amino transferase; ALT= alanine amino transferase

4. CONCLUSION

It can be concluded from the result of this study that inclusion of *Phyllanthus amarus* leaf meal in the diets of broiler finishers do not pose any health hazards to the haematological parameters and serum biochemical profile of the broiler finishers at the levels of inclusion. However, antibacterial properties of *Phyllanthus amarus* should further be verified vis a vis oxytetracycline, so as to be able to recommend it to replace oxytetracycline for antibacterial properties in chickens.

ETHICAL APPROVAL

All authors hereby declare that "Principles of laboratory animal care" (NIH publication No. 85-23, revised 1985) were followed, as well as specific national laws where applicable. All experiments have been examined and approved by the appropriate ethics committee.

COMPETING INTERESTS

Authors have declared that no competing interests exist.

REFERENCES

1. Manning L, Barries RN, Chadd SA. Trends in the global poultry meat supply chain. British Food Journal. 2007;109(5):332-342.
2. Mellon S. Alternatives to antibiotics. Pig Progress. 2000;16:18-21.
3. Ogbuewu IP, Okoli IC, Iloeje MU. Evaluation of toxicological effect of leaf meal on an ethnomedicinal plant-neem on blood chemistry of puberal chinchilla rabbit does. Reports and Opinion. 2010;2(2):29-34.
4. Esonu BO, Opara MN, Okoli IC, Obikaonu HO, Udedibie C, Iheshiulor OOM. Physiological responses of laying birds to neem (*Azadirachta indica*) leaf meal based diets, body weight, organ characteristics and haematology. Online J. Hlth. Allied Sci. Available:http://www.ojhas.org/issue18/2006-2-4.htm
5. Esonu BO, Iheukwumere FC, Emenalom OO, Uchegbu MC and Etuk EB: Performance, nutrient utilization and organ characteristics of broiler fed *Microdesmis pubera* leaf meal. Livestock Research for Rural Development. 2002;14(6).
6. James DB, Owolabi OA, Elebo N, Hassan S, Odemene L. Glucose tolerance test and some biochemical effect of *Phallanthus amarus* aqueous extracts on normaglycemic albino rats. African Journal of Biotechnology. 2009;8(8):1637-1642
7. Taylor L. Technical data report for chanca piedra "stone breaker" (*Phyllanthus niruri*) Preprinted from Herbal Secrets of the Rainforest, 2nd edition. Sage Press, Inc; 2003. Available:www.rain-tree.com/chanca-techreport.pdf
8. Huang Phien; 2009.

Available:http://nampt.com/Kinh-te/Lam-giau-nho-trong-cay-hiem-Diep-ha-chau-cay-dai-t/Default.aspx

9. Sunddaresan NR, Thirumurugan R, Jayakumar S, Purushothaman MR. Protective effectiveness of *Phyllanthus niruri* against short term experimental aflatoxicosis in broiler chicken. Indian Journal of Poultry Science. 2007; 42(2):153-156.

10. Do Tat Loi. Nhung cay thuoc va vi thuoc Viet Nam: Medicine Publisher, Vietnam; 2004.
Available:http://duochanoi.com/diendan/showthread.php?t=2927

11. Nguyen HP, Brian OP and Nguyen QT. Detoxifying effects of a commercial additive and *Phyllanthus amarus* extracts in pigs fed fumonisins-contaminated feed. Livestock Research for Rural Development. 2012;24(6).

12. Kausik B, Ishita C, Ranajil KB, Uday B. Biological activities and medicinal properties of neem *(Azadirachta indica)*. Current Science. 2002;82(11):1336-45

13. Akpan MJ, Enyenihi GE, Obasi OL, Solomon IP, Udedibe ABI. Effects of dietary neem leaf extract on the performance of laying hens. Proceedings of 33rd Annual Conference, NSAP, Nigeria. 2008;396-98.

14. Gafar MK, Itodo AU, Senchi DS. Nutritive and anti-nutritive composition of chanca piedra (stone breaker). Food and Public Health. 2012;2(2):21-27.

15. Mitruka BM, Rawnsly HM. Clinical, biochemical and haematological reference values in normal experimental animals. Masson Publishing, Inc. USA; 1977.

16. AOAC: Official Methods of Analysis. Association of Official Analytical Chemists. 15th Ed. K Helrick. Arlington; 1990.

17. Steel RG, Torrie JHT. Principles and procedures of statistics. An Approach, 2nd ed. McGraw-Hill Book Co. Inc.; 1980.

18. Khan TA, Zafar F. Haematological study in response to varying doses of estrogen in broiler chicken. International Journal of Poultry Science. 2005;10:748-751.

19. Madubuike FN, Ekenyen BU. Haematology and serum biochemistry characteristics of broiler chicks fed varying dietary levels of *Ipomoea asarifolia* leaf meal. Int. J. Poult. Sci. 2006;5:9-12.

20. Adedapo AA, Mogbojuri OM, Emikpe BO. Safety evaluations of the aqueous extract of the leaves of *Moringa oleifera* in rats. Journal of Medicinal Plants Research. 2009;3(8):586-591.
Available:http//:www.academicjournals.org/JMPR

21. Bush BM. Interpretation of laboratory results for small animal clinicians. Blackwell Scientific Publications, Oxford. 1991;11:25-34.

Effect of Process Variables on Survival of Bacteria in Probiotics Enriched Pomegranate Juice

Mina Khanbagi Dogahe[1*], Kianoush Khosravi-Darani[2*], Azadeh Tofighi[1], Mehrnaz Dadgar[1] and Amir Mohammad Mortazavian[3]

[1]Department of Food Science and Technology, Varamin Branch, Islamic Azad University, Varamin, Tehran, Iran.
[2]Research Department of Food Technology, National Nutrition and Food Technology Research Institute, Faculty of Nutrition Sciences and Food Technology, Shahid Beheshti University of Medical Sciences, P.O. Box: 193954741, Tehran, Iran.
[3]Department of Food Sciences and Technology, National Nutrition and Food Technology Research Institute, Faculty of Nutrition Sciences and Food Technology, Shahid Beheshti University of Medical Sciences, Tehran, Iran.

Authors' contributions

This work was carried out in collaboration between all authors. All authors read and approved the final manuscript.

Editor(s):
(1) Yan Juan, Sichuan Agricultural University, Ya'an, China.
Reviewers:
(1) Anonymous, Italy.
(2) Beatriz de Cássia Martins Salomão, Dept. of Chemical Engineering, Federal University of Rio Grande do Norte, Brazil.
(3) Anonymous, Egypt.

ABSTRACT

Aims: The consumption of foods and beverages containing probiotic microorganisms is a growing, global consumer trend. In this research, production of probiotic pomegranate juice containing *Lactobacillus plantarum* and *Lactobacillus delbrueckii* was studied.
Study Design: Plackett-Burman statistical design was used to evaluate the impact of eleven process variables on the viability of both probiotics. Impact of incorporation of grape juice, tomato juice and pomegranate peel extract as well as phenolic compounds and vitamins have been investigated.
Place and Duration of Study: *Department of Food Science and Technology,* Varamin Branch, Islamic Azad University, Varamin, Tehran, Iran, between Sep 2012 and July 2013.
Methodology: Pomegranate juices were inoculated with probiotic bacteria and their survival was

Corresponding author: Email: k.khosravi@sbmu.ac.ir, kiankh@yahoo.com evaluated every week by pure plate method. The effect of 11 variables (in two levels) on survival of bacteria by the statistical design of Plackett-Burman was evaluated. For this purpose, 12 treatments in triplicate by the Minitab (version = 11.0) software at significant levels α= .01 were analyzed.

Results: The highest survival rate of L. plantarum (4.74 ×10^6 CFU/mL) and L. delbrueckii (4 ×10^6 CFU/mL) was obtained by 10% v/v inoculation of a 48 h inoculum culture in MRS broth medium to enriched pomegranate juice (10% v/v Grape juice, 5% v/v tomato juice, 0.1% v/v pomegranate peel extract and 2.0 g.L glucose) which was inoculated in anaerobic condition for 72 h at 37°C and kept for 2 weeks at environment temperature. Sensory evaluation shows the probiotic juice was accepted by consumers with no significant difference in comparison to control in terms of taste, odour and overall acceptability (P>.05).

Conclusion: The results of this study suggest that grape, tomato juices and pomegranate peel extract exert a protective effect on L. plantarum and L. delbrueckii viability under acidic condition of pomegranate juice and storage time, which was associated with the chemical composition of them. This study indicates that develop of probiotic pomegranate juices with acceptable viability and stability of the probiotic is possible.

Keywords: Probiotic; pomegranate juice; survival; process variables; plackett-burman statistical design.

1. INTRODUCTION

The consumption of foods and beverages containing probiotic microorganisms is a growing, global consumer trend [1]. The term probiotic has technically defined as live microorganism which following uptake in certain numbers exerts health benefits beyond inherent general nutrition [2,3]. Functional food with probiotics, prebiotics and fibers has been available for consumers for years. Prebiotics are non digestible food ingredients that beneficially affect the host by selectively stimulating the growth and/or activity of one or a limited number of bacteria in the colon and thus improve host health [4]. Synbiotic food products containing probiotics with prebiotics or fibers are gaining more interest into the market [5]. Research has shown that addition of probiotics to food provides several health benefits including reduction in the level of serum cholesterol, improved gastrointestinal function, enhanced immune system and lower risk of colon cancer [6-9]. Traditionally, probiotics have been added to yogurt and other fermented dairy products, but lactose intolerance and the cholesterol content are two drawbacks related to consumption of milk products. In recent years, consumers' demand for nondairy based probiotic products has increased [10].

Fruit juice has been suggested as a novel, appropriate medium for fortification with probiotic cultures because it is already positioned as a healthy food product. It is consumed frequently by a large percentage of the consumer population [11]. Fruits and vegetables are rich in functional food components such as minerals, vitamins, dietary fibers and antioxidants. Furthermore, they do not contain any dairy allergens that might prevent usage by certain segments of the population [12]. It has been suggested that fruit juices could serve as suitable media for cultivating probiotic bacteria [13].

Pomegranate and its products, including juice, tea, wine and extracts are widely consumed and recognized for their health benefits. For instance, commercially manufactured pomegranate juice has a higher antioxidant activity than red wine and green tea [14]. The fresh juice contains 85.4% water and considerable amounts of total soluble solids (TSS), total sugars, reducing sugars, anthocyanins, phenolics, ascorbic acid and proteins and has been reported to be a rich source of antioxidants [15].

About half of the total dry matter of tomato consists of the reducing sugars, glucose and fructose, about 10% is organic acid and about 1% is skin and seeds, with the remainder being alcohol, insoluble solids (cellulose, pectins, hemicellulose and proteins), mineral (mainly potassium), pigments, vitamins and lipids. Glutamic acid is the principal amino acid found in tomato [16]. The organic acids content of tomato is responsible for a pH between 4.0 and 4.6 [17]. Tomato juice is a rich source of simple sugars, minerals and vitamins and its use has often been suggested to manufacture acceptable acidophilus milk products [18].

Grape pomace is a natural product rich in dietary fiber and polyphenols. An increasing interest on dietary phenolic compounds intake has been observed due to their well known antioxidant properties and health benefits e.g. significant reduction of cardiovascular disease risk [19]. The lactic acid bacteria (LAB) occurs naturally on grapes and their ability to grow in grape juice and wine is well documented [20].

Over centuries, fermentation has been used to improve the quality and the flavour of cereals, fruits, vegetables, legumes and meat. Nonetheless, growing body of evidence documenting the beneficial health effects of probiotics, there is a lack of research on the application of probiotics in product development [21]. Different studies have been carried out to explore the suitability of fruit juices such as tomato, beet, cabbage and juices as raw material for production of fruit drinks. *Lactobacillus plantarum*, *Lactobacillus acidophilus* and *Lactobacillus casei* have been employed as probiotic bacteria culture. Results have indicated that all the strains are capable to growth in the fruit juices mentioned and, as a result, the microbial population increase after 48 h of fermentation. Moreover, *L. plantarum*, *L. acidophilus* and *Lactobacillus delbrueckii* have shown to be resistant to high acidic and low pH conditions during storage periods at 4°C [22-24]. The suitability of pomegranate juice as raw material for production of probiotic fruit juice has been studied. *L. plantarum*, *L. delbrueckii*, *Lactobacillus paracasei*, *L. acidophilus* were examined. *L. plantarum*, *L. delbrueckii* showed higher viability during the storage time. Pomegranate juice was proved to be a suitable media for production of a fermented probiotic drink [25]. The survival of probiotic in a model fruit juice system has been studied. The model juice containing vitamins, grape extract and green tea extract showed better survival of probiotic bacteria [26]. Table 4 summarizes kind and status of the aforementioned microorganism.

In the present work, the stability of *L. plantarum* and *L. delbrueckii* cells in pomegranate juice was studied. Pomegranate was selected as an important source of bioactive compounds with high antioxidant activity. Its peel extract has also high antioxidant activity due to high level of phenolics compounds. Fresh probiotic biomass was used instead of lyophilized bacteria. The effect of 11 variables (selected by pre-experience and literature review) on survival of bacteria was investigated by Plackett-Burman design (PBD).

Grape juice, tomato juice and pomegranate peel extract were selected as antioxidant natural sources of phenolic compounds and vitamins.

2. MATERIALS AND METHODS

2.1 Starter Culture

The microorganisms used in the present study were as following: *L. plantarum* PTCC (Persian Type Culture Collection) NO: 1745 (isolated from pickled cabbage) and *L. delbrueckii*: PTCC NO: 1333, (isolated from intestine of adult) which were purchased from Iranian Research Organization for Science and Technology (IROST). The strains were maintained at 4°C and subcultures monthly on MRS agar (Merc, Germany) plate to maintain freshness. Also strains were precultured weekly in MRS broth (Merc, Germany) at 37°C for 24 h (stationary phase) under anaerobic conditions [27].

Number of precultured bacteria was measured by spectrophotometer and microscope to draw calibration curve of number of viable cells. First precultured bacteria was prepared by addition of 1.0 mL of the incubated tubes (with estimated viable cell of *L. plantarum* from 6 to 6.68 log CFU/mL and *L. delbrueckii* from 6 to 6.6 log CFU/mL) to 9.0 and 8.0 mL of sterilized saline solution (0.9% w/v), before use as 10^{-1} and 2×10^{-1} dilution. Then, serial dilutions of 10^{-2} to 10^{-5} were prepared as mentioned previously and the absorbance at 600 nm was measured with spectrophotometer (CT chromtech, Taiwan). The direct colony count method was used to determine cell viability too. Cell concentrations were measured by staining and microscope direct counting and calibration curve was plotted.

2.2 Preparation of Product

Pomegranates, grapes and tomato juice were prepared manually then pasteurized for 5 min at 80°C by water bath and stored at refrigerator temperature. Pomegranates juices were weighed in 40.0 mL packages. Glucose, fructose, citric acid, pomegranate peel extract, grapes juice and tomato juice were added to the pomegranates juices at two levels according to statistical design of Plackett-Burman. Precultured bacteria (cultured in MRS broth medium for 24 and 48 h at 37°C) were centrifuged (5000×g, 10 min, 4°C), the pellet was washed twice with sterile saline (0.9% w/v) and the cells resuspended in sterile saline (0.9% w/v) [28,29].

The centrifuged cells were inoculated to the pasteurized pomegranates juices (10% or 20% v/v) and incubated at 37°C for 3 and 4 days. Products were kept at 4°C and 20°C for 1 and 2 weeks. Fig. 1 shows a schematic diagram of fortified pomegranates juice (pomegranates juice with grape and tomato juice, pomegranate peel extract, citric acid, fructose and glucose which was inoculated with probiotic bacteria) production (Fig. 1).

2.3 Bacterial Enumeration

The pure plate method was used to determine viable cell count. Samples (products) were diluted ($10^{-4} - 10^{-6}$) with sterile saline (0.9% w/v) solution. The plates were incubated at 37°C for 48 h and those showing individual colonies in the plates were counted. Viable cell counts were calculated as colony forming units per milliliter (CFU/mL).

2.4 Sensory Evaluation of Product

The taste, odour and overall acceptability of the simple pomegranates juice (not enriched) and the samples which have the highest viable cells count were judged by 30 untrained panellists. A nine points structured hedonic scale was used during a sequential presentation of samples, with one = disliked very much and nine = liked very much. The results were statistically analyzed using an ANOVA and Duncan method (.01<P<.05) (27).

2.5 Statistical Analysis

The statistical design of Plackett-Burman was used to figure out the optimum conditions for producing probiotic pomegranates juice with *L. plantarum* and *L. delbrueckii*. Each culture was used separately in a single product. The effect of each factor was evaluated (α = .01).

The basic equation set up for the design was as follows. The coefficients for the eleven variables were determined by:

$$A_i = (1/N) \sum_0^n X_i \cdot K_i$$

where A_i = coefficient values, X_i = experimental yield, K_i = codeed value of each variable corresponding to the respective experimental

yield X_i and N = number of experiments. For Predicted yield is given by:

$$Y_i = \sum_{i=0}^N A_i \cdot K_i$$

for i = 0, a dummy level of +1 was used and the coefficient obtained was called A_0. The standard error was determined as the sum of the squares of the difference between the experimental and predicted yield for each run. The estimated error is given by the following equation:

$$S_b = \sqrt{S_e^2 / N}$$

The student's t test was performed to estimate the significance of each variable employed (t value =coefficient/S_b). Since the experiments were designed to evaluate the relative effect of each variable on response, a significant level of 0.30 is acceptable [30]. However, the tabulated t value (degree of freedom 10) at P<.1 and P<.15 is equal to 1.2 and 0.69, respectively.

2.5.1 Selection of process variables and range finding

The first screening step is usually identification of the variables which have significant effects on response. With the aim to increase the viability of both probiotics, selection of the factors and their range (in a wide but reasonable numerical range) was conducted based on literature review and also our previous experience (unpublished data). Some changes in the response (yield) were expected for each factor over the selected range [31,32]. The variables to be evaluated are summarized in Table 1. Selected experimental variables and a Plackett-Burman design (PBD) for conducting 12 experimental trials were shown in Table 2. The elements + (high level) and - (low level) represent the two different levels of the independent variables examined.

2.6 Survival in the Simulated Gastro-Intestinal Conditions

After 30 days of storage of probiotic pomegranate juice at 4°C, survival of probiotic bacteria of samples exposed to simulated gastro-intestinal stresses were compared to the control (a freshly prepared culture before storage) [33]. MRS broth (this type of bacterial growth medium is so-named by its inventors: De Man, Rogosa and Sharpe) was prepared and used as the base medium (BM) to evaluate influence of pancreatic enzymes, pH and bile salts on cell viability. After sterilization by autoclave at 121°C for 15 min, BM was adjusted at pH 2.0 by 6.0 N HCl solution for evaluation of stability at pH 2.0.

To prepare BM containing bile salt, 0.3% of bovine Oxgall was added to BM, then pH of the medium was adjusted to 6.0.

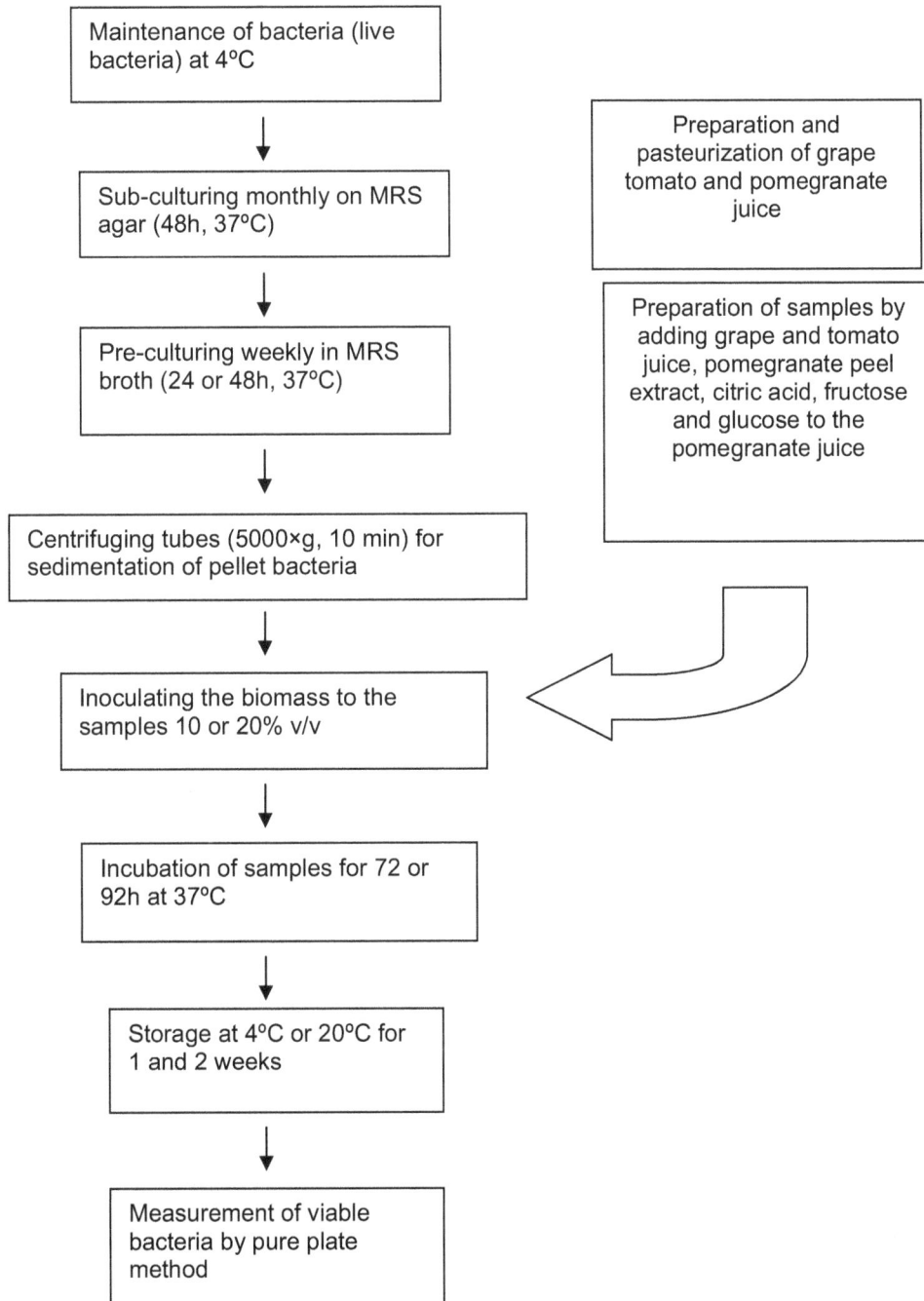

Fig. 1. Schematic diagram of enriched pomegranates juice production

Table 1. The independent variables to be evaluated and their importance in production of probiotic pomegranate juice

Independent variables	Range finding	Low level (-)	High level (+)	Suitable amount	Role	Reference
A: Grape juice content	5 and 10 (%v/v)	5	10	10 [a,b]	Positive effect of the total antioxidant capacity and phenolic content on bacteria viability	[35]
B: Tomato juice content	5 and 10 (%v/v)	5	10	10 [a,b]	Positive effect of simple sugars (glucose, fructose) and minerals (Mn, Mg) on bacteria viability	[25]
C: Pomegranate peel extract content	0.01 and 0.1 (%v/v)	0.01	0.1	0.1 [a,b]	pomegranate peel extract is a good source of the total antioxidant capacity and phenolic content	[46]
D: Fructose concentration	0 and 1 (g/L)	0	1	1 [b], 0 [a]	Fructose is one of the sugar that often consumed by probiotic bacteria	[25]
E: Acetic acid concentration	0 and 1 (g/L)	0	1	0 [a,b]	As the main source of carbon used by lactic acid bacteria	[25]
F: Inoculums content	10 and 20(%v/v)	10	20	Not significant	Reaching to high cell concentration and inoculums in less percent	[25]
G: Inoculum age	24 and 48 (h)	24	48	48 [a], 24 [b]	Presence of glucose and fructose in the medium led to longer lag phase	[37]
H: Storage temperature	4 and 20 (°C)	4	20	20 [a,b]	4 as refrigerator and 20 as room temperature	[36]
I: Storage time	1 and 2 (week)	1	2	2 [a,b]	–	–
J: Fermentation time	72 and 92 (h)	72	92	72 [a,b]	The best fermentation time for probiotic bacteria was 72h or more of its in pomegranate juice	[25]
K: Glucose concentration	0 and 1 (g/L)	0	1	1 [a,b]	Glucose increased the survival of lactobacilli in acidic conditions	[25]

[a]: Suitable range for *L. delbrueckii* [b]: Suitable range for *L. plantarum*

After sterilization of the solution by filtration and reaching temperature to 37°C, medium was inoculated with 10% (v/v) of the probiotic containing pomegranate juice and incubated for 2 h. Cell counts were conducted on MRS agar and the control was base medium at pH 6.0. The "fresh culture" control was prepared by centrifuging a freshly-grown MRS culture at 3000×g for 15 min and adding the pomegranate juice to the cell pellet in order to obtain approximately 10^7 CFU/mL.

Pancreatin solution (Sigma) was centrifuged at 2000×g for 15 min at 4°C and sterilized by microfiltration (0.22 μm pore), then 50 mg/mL of this solution was added to BM. Finally, 380 μL of the enzyme solution was aseptically added to 10.0 mL of the BM to a final pancreatin concentration of 1.9 mg/mL and pH 6.0.

3. RESULTS AND DISCUSSION

3.1 Stability in Pomegranate Juice

After the addition of ingredients to pomegranate juice and inoculation, the cell count was measured at the beginning of the storage time (time=0) as well as 1 and 2 weeks after storage at 4°C and 20°C.

Table 2. Studding eleven factors in probiotic juice production by adding *L. plantarum* and *L. delbrueckii* separately using the PBD statistical design

Run	Coded setting for factors											Response[a] (CFU/mL)	
No	A	B	C	D	E	F	G	H	I	J	K	*L. plantarum*	*L. delbrueckii*
1	+	-	+	-	-	-	+	+	+	-	+	$4.74 \times 10^6 \pm 0.11$	$4 \times 10^6 \pm 0.2$
2	+	+	-	+	-	-	-	+	+	+	-	$4.15 \times 10^6 \pm 0.13$	$3.40 \times 10^6 \pm 0.14$
3	-	+	+	-	+	-	-	-	+	+	+	$3.80 \times 10^6 \pm 0.11$	$3.10 \times 10^6 \pm 0.07$
4	+	-	+	+	-	+	-	-	-	+	+	$3.60 \times 10^6 \pm 0.03$	$2.30 \times 10^6 \pm 0.18$
5	+	+	-	+	+	-	+	-	-	-	+	$2.70 \times 10^6 \pm 0.05$	$2.50 \times 10^6 \pm 0.08$
6	+	+	+	-	+	+	-	+	-	-	-	$3.40 \times 10^6 \pm 0.17$	$2.90 \times 10^6 \pm 0.03$
7	-	+	+	+	-	+	+	-	+	-	-	$3.90 \times 10^6 \pm 0.03$	$3.60 \times 10^6 \pm 0.11$
8	-	-	+	+	+	-	+	+	-	+	-	$2.20 \times 10^6 \pm 0.06$	$1.80 \times 10^6 \pm 0.07$
9	-	-	-	+	+	+	-	+	+	-	+	$3.20 \times 10^6 \pm 0.09$	$2.40 \times 10^6 \pm 0.06$
10	+	-	-	-	+	+	+	-	+	+	-	$2.90 \times 10^6 \pm 0.10$	$2.90 \times 10^6 \pm 0.21$
11	-	+	-	-	-	+	+	+	-	+	+	$2.50 \times 10^6 \pm 0.03$	$2.70 \times 10^6 \pm 0.12$
12	-	-	-	-	-	-	-	-	-	-	-	$2.02 \times 10^6 \pm 0.10$	$1.90 \times 10^6 \pm 0.06$

[a]: *response shows as mean±standard deviation for 3 replications*

The cell viability in all samples were measured at the beginning of storage (time=0). To study effect of time on survival of bacterial cell, cell count was also conducted after 1 and 2 weeks (time=1 and 2). As Fig. 2 a shows, in some trials the number of viable cells was changed during the storage time. The maximum and minimum changes of *L. plantarum* were 0.16 and 0.02 log CFU/mL respectively and mean of changes was 0.07±0.001 log CFU/mL. Also, the maximum and minimum changes of L. delbrueckii were 0.2 and 0 log CFU/mL respectively and mean of changes was 0.09±0.002 log CFU/mL.

In the trials containing *L. plantarum*, at the beginning of storage, the highest and lowest numbers of live bacteria were respectively belonged to the test number 9 and 12. The highest numbers of viable bacteria in the seventh and fourteenth day of storage were respectively belonged to the test number 4 and 1. In other words, in sample 1 the highest numbers of viable bacteria was achieved after 2 weeks of storage. In general, survival of *L. plantarum* increased after 2 weeks so it could adapt with environment (Fig. 2a).

In the case of *L. delbrueckii* (Fig. 2b), the highest numbers of living bacteria were belonged to the test number 1 at the beginning and also after 2 weeks of storage.

3.2 Statistical Evaluation of Cell Viability

Plackett-Burman statistical design was used to evaluate the impact of 11 process variables (in 2 levels) on the viability of probiotics. The results are shown in Figs. 3 and 4 for *L. plantarum* and *L. delbrueckii*. All variables, except for the inoculums concentration, were conceded as significant effective variables on viability of both bacteria. Storage time, pomegranate peel extract, grape juice and acetic acid concentration have been found more effective than other variables on viability of *L. plantarum* (Fig. 3a). In the trials inoculated by *L. delbrueckii* storage time, tomato juice, grape juice and acetic acid concentration have been found more effective than other variables (Fig. 3b).

The high level of grape juice, tomato juice and pomegranate peel extract concentration, because of their total antioxidant capacity, minerals and phenolic contents leads to a positive effect on viability of both bacteria. Glucose and fructose are metabolized by *L. plantarum* and *L. delbrueckii*. As it can be concluded from Fig. 4, high concentration of glucose and fructose were found appropriate for *L. plantarum* but for viability of *L. delbrueckii* high level of glucose in the absence of fructose was more suitable. It seems that, excessive increase of sugars causes a decrease in survival of *L. delbrueckii*.

Due to the significant decrease in pH after the addition of citric acid, reducing of the survival of bacteria can be justified. Thus, absence of citric acid in the product was found suitable for both bacteria. Suitable age of inoculums for *L. plantarum* was 24 h. Studies have shown that log phase of growth has elongated in the presence of glucose and stationary phase start after 24 h [28]. So, the results indicated that inoculation with 24 h cultured cells, was more appropriate than 48 h for *L. plantarum*. In another study, it has been shown that the growth curve of *L. delbrueckii* was entered to the stationary phase after 25 h [25]. Also in the presence of glucose log phase was elongated, so, in this study, inoculums age of 48 h seems more suitable than 24 h for *L. delbrueckii*. Microbial population of *L. plantarum* and *L. delbrueckii* decreased during fermentation time (92 h) at 37°C but fermentation time of 72 h has been shown to be more effective. In other words suitable fermentation time was 72 h for both bacteria. Storage of product at room temperature (20°C) for 2 weeks has shown positive effect on both bacteria but at 4°C growth of bacteria was slower. Taste, odour and overall acceptability of probiotic pomegranate juice were examined. Samples which had the highest viable cell were compared with the control sample (pomegranate juice without any additions) by nine point structured hedonic scale (Table 3). The results were examined by ANOVA and F test. No significant difference among between samples and control in terms of taste, odour and overall acceptability were observed (.01<*P*<.05).

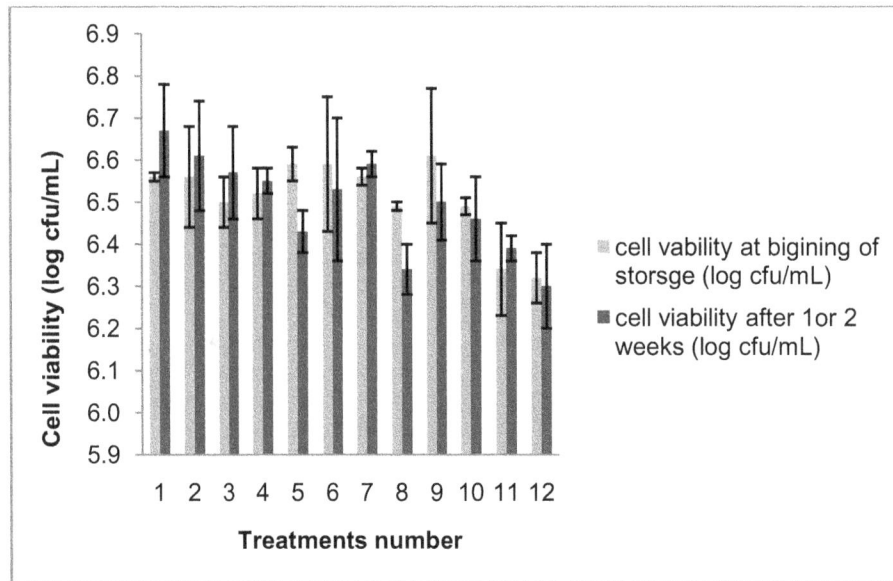

(a)

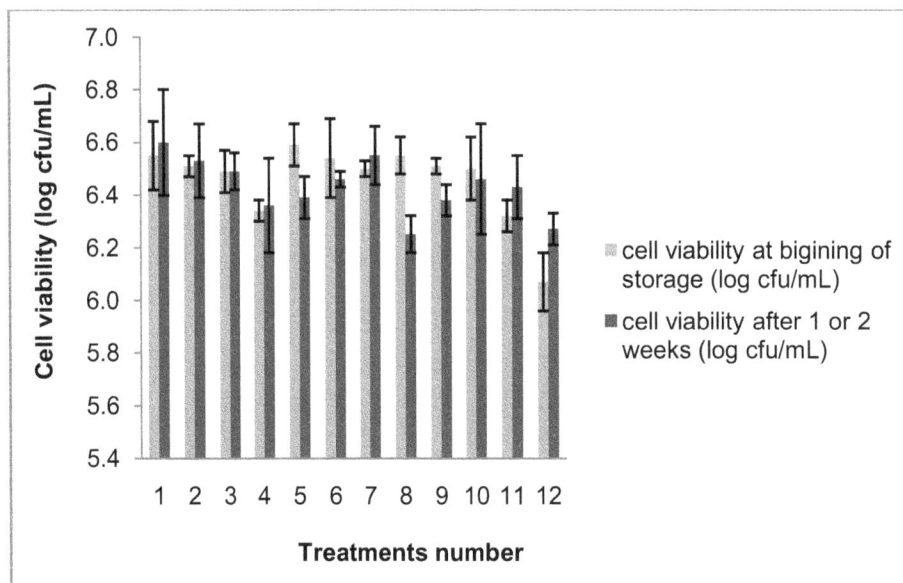

(b)

Fig. 2. Stability of fresh *L. plantarum* (a) and *L. delbrueckii* (b) in fortified pomegranate juice during storage time in 12 treatments different conditions according to the plackett-burmann design

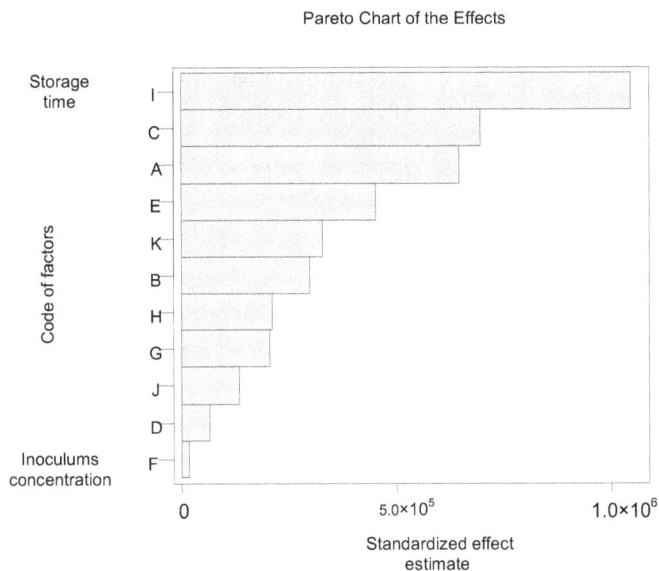

(a)

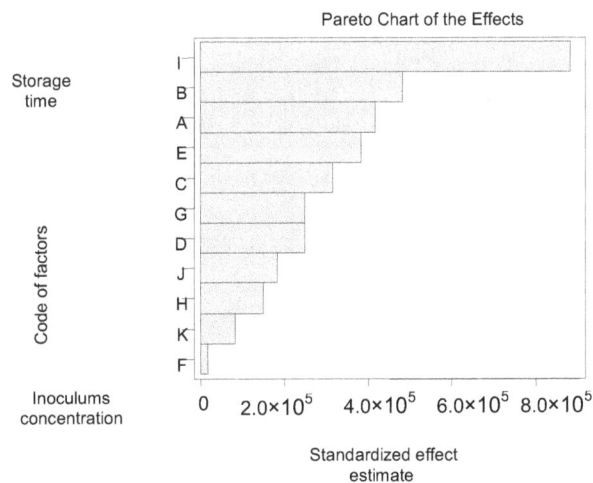

(b)

Fig. 3. Pareto diagram to evaluate the effect of eleven variables (A: grape juice content, B: tomato juice content, C: pomegranate peel extract content, D: fructose concentration, E: acetic acid concentration, F: inoculums content, G: inoculum age, H: storage temperature, I: storage time, J: fermentation time, K: glucose concentration) on survival of *L. plantarum* (a) and *L delbrueckii* (b) inoculated in pomegranate juice during storage time (significant level: α=.01)

As results show the best condition for survival of probiotic bacteria was achieved when bacteria incubated in MRS broth for 48 h and inoculated in pomegranate juice contains grape juice (10% v/v), tomato juice (5% v/v), pomegranate peel extract (0.1% v/v) and glucose (1.0 g/L) and incubated for 72 h at 37°C in anaerobic condition then products were stored at 20°C for 2 weeks.

The highest survival number of *L. plantarum* and *L. delbrueckii* were about 4.74×10^6 and 4×10^6 CFU/mL, respectively. Grape juice contains high rate of antioxidant, minerals and phenolic compounds which are effective on growth and viability of probiotic bacteria [34]. Vitamins, pigments and antioxidant substances in the tomato made the medium more suitable for fermentation. This result is similar with previous studies [22]. The growth and survival of probiotic bacteria also were stimulated by presence of phenolic compounds found in pomegranate peel extract. phenolic compounds not only inhibit the growth and activity of spoilage microorganisms, but also have stimulation effect on the growth of probiotic bacteria [26,37]. Using of appropriate rate of pomegranate peel in the juice leads to increase survival of bacteria without creation of undesirable flavour. Glucose was used as a carbon source by bacteria so, it had influencing role on growth and viability of bacteria. The probiotic pomegranate juice did not show any significant difference with the control and was acceptable for consumers.

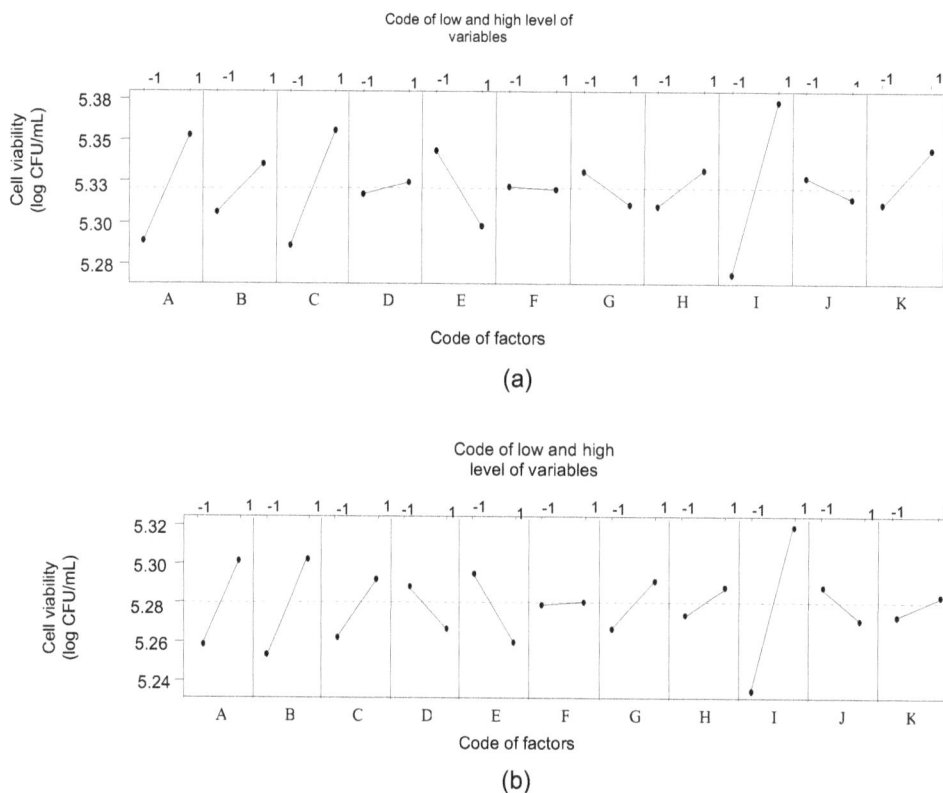

(a)

(b)

Fig. 4. The main effect of different level variables (A: grape juice content, B: tomato juice content, C: pomegranate peel extract content, D: fructose concentration, E: acetic acid concentration, F: inoculums content, G: inoculum age, H: storage temperature, I: storage time, J: fermentation time, K: glucose concentration) in plackett-burmann design on survival of *L. plantarum* (a) and *L delbrueckii* (b) inoculated in pomegranate juice during storage

Table 3. Analysis result of taste, smell and acceptances for the control sample, sample number 1 for both bacteria (as best products that have the highest live bacteria)

Sample	Sensory indicators			Total score
	Taste	Smell	Acceptance	
Control	7.1	5.8	6.4	19.3
Trial no. 1 [a]	5.8	4.9	5.3	16
Trial no. 1 [b]	6.2	5.1	5.5	16.8

(a): Sample that contains the highest number of live *L. plantarum*;
(b): Sample that contains the highest number of live *L. delbrueckii*

Table 4. Probiotics used in different types of juices in liquid fermentation condition

Probiotic bacteria	Inoculum size (log CFU/mL)	Product	Survival rate (CFU/mL)	Consideration	Reference
L. plantarum	1.64	Tomato juice	10^6	Viability in product	[22]
L. acidophilus			10^9		
L. casei			10^8		
L. delbrueckii			10^8		
L. plantarum	5.20	Beet juice	10^6	Viability in product	[23]
L. acidophilus			10^4		
L. casei			10^7		
L. delbrueckii			10^7		
L. delbrueckii	2.9	Cabbage juice	10^7	Viability in product	[24]
L. plantarum			10^5		
L. casei			10^5		
L. delbrueckii	0.049	Pomegranate juice	10^5	Viability in product	[25]
L. plantarum					
L. casei					
L. acidophilus					

3.3 Viability during Simulated Gastro-intestinal Stresses

Some strains were selected to study their resistance to simulated gastro-intestinal (GI) conditions, mainly based on variable stability during storage. Indeed, *Lactobacillus rhamnosus* was the most stable cultures. Furthermore, *Lactobacillus reuteri* was of interest because it is considered to be of the few true indigenous *Lactobacillus* species in humans [38] while *L. plantarum* is rather associated with the fermentation of vegetal substrates [39]. None of the strains were significantly affected by the incubation in presence of 0.3% bile salts or the pancreatic enzymes.

Storage for 30 days in the fruit juice blend did not affect the strains sensitivity to these compounds. In comparison to control treatment (base medium at pH 6.0), viability loss of *L. plantarum* after treatment for acid (pH 2), bile (0.3%) and pancreatic enzymes was 1.8, 0.7 and 0 log CFU/mL. Also viability loss of *L. delbrueckii.* after treatment for acid (pH 2), bile (0.3%) and pancreatic enzymes was 1.6, 0.2 and 0 log CFU/mL. In probiotic selection, tolerance to the environment of the small intestine is also thought to be of importance since the acid-sensitive strains can be buffered through the stomach [40,41]. The results of this investigation show that 30 days storage in the fruit drink would not affect sensitivity of probiotics to bile or pancreatic enzymes. The previous reports indicate that a low final pH during bacterial growth induces an acid tolerance response [42,43]. The induction of pH stress may cause protection against acid, heat, osmotic or oxidative shocks [44]. The

opposite observation was reported, when strong viability losses were observed by exposing at pH 2.0 for 2 h at 37°C [45].

Existence of direct relationship between stability during storage in the juice and simulated GI condition was examined. The correlation (R^2 = 0.53) was not statistically significant (P=.2) due to the limited number of experimental data (only 2 strains). It has been shown that glucose enhanced the survival of lactobacilli in acidic conditions [47]. Therefore, in order to prevent viability loses during storage and in the GI stress tests, incorporation of a carbohydrate can be useful. This being stated, this classical simulated GI acid procedure at pH levels between 1.5 and 2.5 probably overestimates actual gastric viability losses due to the buffering ability of certain foods [48].

It should be mentioned that in all fractional factorial designs like PBD [49] and Taguchi Design [50], the better level of each variable can be defined to achieve the better response. In this study, PBD was applied for producing probiotic pomegranates juice with *L. plantarum* and *L. delbrueckii*. The study can not be considered as optimization process, but it is only a screening design to evaluate the effect of independent factor and finding the more suitable level (between two selected levels) of independent variable.

4. CONCLUSION

The present study indicated that some mix juice preparations have the potential in technological applications in protecting probiotic viability and stability during processing and storage of fruit juices. The stored cultures seem to be less resistant than fresh cells to an acid challenge as would occur in the stomach following consumption of the products. This was observed with probiotic bacteria stored in a fruit juice, but raises a concern with respect to other acid foods such as pomegranate juice. The results presented here suggest that grape, tomato juices and pomegranate peel extract exert a protective effect on *L. plantarum* and *L. delbrueckii* viability under acidic condition of pomegranate juice and storage time, which was associated with the chemical composition of the aforementioned ingredients. Despite the low percentage of pomegranate peel extract (0.1% or 0.01% v/v), had a significant effect in this regard. These effects could be mainly attributed to the presence of antioxidant, minerals and phenolic compounds present in the pomegranate peel extract. These results indicate that it is possible to develop probiotic fruit juices where the juices maintain the viability and stability of the probiotic. It remains to be determined the effect of storage on the functional properties of probiotics as well as influence of the age of the stored cultures on clinical health benefits.

COMPETING INTERESTS

Authors have declared that no competing interests exist.

REFERENCES

1. Verbeke W. Consumer acceptance of functional foods: Sociodemographic cognitive and attitudinal determinants. Food Qual Prefer. 2005;16(1):45-57.
2. FAO/WHO. Health and nutritional properties of probiotics in food including powder milk with live lactic acid bacteria. Cordoba, Argentina: Food and Agriculture Organization of the United Nations and World Health Organization Expert Consultation Report; 2001.
3. Guarner F, Schaafsma GJ. Probiotics. Int J Food Microbiol. 1998;39:237-238.
4. Gibson GR, Roberfroid MB. Dietary modulation of the human colonic microbiota: Introducing the concept of prebiotics. J Nutr. 1995;125:1401-1412.
5. Sloan AE. The top ten functional food trends. Food Technol. 2000;54:33-62.
6. Berner LO, Donnell J. Functional foods and health claims legislation: Applications to dairy foods. Int Dairy J. 1998;8:355-362.
7. Mcnaught CE, Macnie J. Probiotics in clinical practice: A critical review of the evidence. Nutr Res. 2001;21:343-353.
8. Rafter J. Probiotics and colon cancer. Best Pract Res Cl Ga. 2003;17(5):849-859.
9. Saarela M, La"hteena"ki L, Crittenden R, Salminen S, Mattila Sandholm T. Gut bacteria and health foods—the European perspective. Int J Food Microbiol. 2002;78:99-117.
10. Shah NP. Functional foods from probiotics and prebiotics. Food Technol. 2001;55(11):46–53.
11. Tuorila H, Gardello AV. Consumer response to an off flavor in juice in the presence of specific health claims. Food Qual Prefer. 2002;13:561-569.
12. Luckow T, Delahunty C. Which juice is healthier? A consumer study of probiotic nondairy juice drinks. Food Qual Prefer. 2004;15:751-759.
13. Mattila Sandholm T, Myllarinen P, Crittenden R, Mogensen G, Fonden R, Saarela M. Technological challenges for future probiotic foods. Int Dairy J. 2002;12:173-182.
14. Gil MI, Tomas Barberan FA, Hess Pierce B, Holcroft DM, Kader AA. Antioxidant activity of pomegranate juice and its relationship with phenolic composition and processing. J Agr Food Chem. 2000;48:4581-4589.
15. Aviram M, Dornfeld L, Kaplan M, Coleman R, Gaitini D, Nitecki S, et al. Pomegranate juice flavonoids inhibit LDL oxidation and cardiovascular diseases: Studies in atherosclerotic mice and in humans. Drugs Exp Cl Res. 2002;28:49-62.
16. Hayes WA, Smith PG, Morris AE. The production and quality of tomato concentrates. Crit Rev Food Sci Nutr. 1998;38(7):537-564.
17. Gutheil RA, Price LG, Swanson BG. pH, acidity and vitamin C content of fresh and canned homegrown Washington tomatoes. J Food Protect. 1980;43:366-369.
18. Babu V, Mital BK, Garg SK. Effect of tomato juice addition on the growth and activity of *Lactobacillus acidophilus*. Int J Food Microbiol. 1992;17:67-71.
19. Hervert Hernández D, Pintado C, Rotger R, Goñi I. Stimulatory role of grape pomace polyphenols on *Lactobacillus acidophilus* growth. Intl J Food Microbiol. 2009;136:119-122.
20. Plessis HW, Dicks LMT, Pretorius IS, Lambrechts MG, du Toit M. Identification of

lactic acid bacteria isolated from South African brandy base wines. Int J Food Microbiol. 2004;91:19-29.

21. Rivera Espinoza Y, Gallardo Navarro Y. Nondairy probiotic products. Food Microbiol. 2010;27:111.

22. Yoon KY, Woodams EE, Hang YD. Probiotication of tomato juice by lactic acid bacteria. J Microbial. 2004;42:315-318.

23. Yoon KY, Woodams EE, Hang YD. Fermentation of beet juice by beneficial lactic acid bacteria. Lebensm Wiss u Technol. 2005;38:73-75.

24. Yoon KY, Woodams EE, Hang YD. Production of cabbage juice by lactic acid bacteria. Bioresource Technol. 2006;97:1427-1430.

25. Mousavi ZE, Mousavi SM, Razavi SH, Emam Djomeh Z, Kiani H. Fermentation of pomegranate juice by probiotic lactic acid bacteria. World J Microbiol Biotechnol; 2010. DOI 10.1007/s1127401004361.

26. Shah NP, Ding WK, Fallourd MJ, Leyer G. Improving the stability of probiotic bacteria in model fruit juice using vitamins and antioxidants. J Food Sci. 2010;75(5):278-282.

27. Guergoletto K, Magnani M, San Martin J, Andrade C, Garcia S. Survival of *Lactobacillus casei* (LC1) adhered to prebiotic vegetal fibers. Innov Food Sci and Emerg Tech. 2010;11:415-421.

28. Charalampopoulos D, Pandiella SS, Webb C. Evaluation of the effect of malt, wheat and barley extracts on the viability of potentially probiotic lactic acid bacteria under acidic conditions. Int J Food Microbiol. 2003;82:133-141.

29. Michida H, Tamalampudi S, Pandiella S, Webb C, Fukuda H, Kondo A. Effect of cereal extracts and cereal fiber on viability of *Lactobacillus plantarum* under gastrointestinal tract conditions. Biochem Eng J. 2006;28:73-78.

30. Khosravi Darani K, Zoghi A. Comparison of pretreatment strategies of sugarcane baggase: Experimental design for citric acid production. Bioresource Technol. 2008;99:6986–6993.

31. Davies L. Efficiency in research, development and production: The statistical design and analysis of chemical experiments. Trends Anal Chem. 1995;4:19-50.

32. Khosravi Darani K, Vasheghani Farahani E, Shojaosadati SA. Application of the Plackett Burman design for the optimization of poly (β-hydroxybutyrate) production by *Ralstonia eutropha*. Iran J Biotechnol. 2003;1(3):155-161.

33. Charteris WP, O'Neill ET, Kelly PM. An in vitro method to determine human gastric transit tolerance among potentially probiotic lactic acid bacteria and bifidobacteria. Irish J Agr Food Res. 1994;33:203.

34. Wang H, Cao G, Prior RL. Total antioxidant capacity of fruits. J Agric Food Chem. 1996;44:701-705.

35. Kneifel W, Jaros D, Erhard F. Microflora and acidification properties of yogurt and yogurt related products fermented with commercially available starter cultures. Int J Food Microbiol. 1993;18:179-189.

36. Bialonska D, Kasimsetty SG, Schreder KK, Ferreira D. The effect of pomegranate *(punica granatum L.)* byproducts and ellagitannins on the growth of human gut bacteria. J Agric Food Chem. 2009;57:8344-8349.

37. Kneifel W, Rajal A, Kulbe KD. *In vitro* behavior of probiotic bacteria in culture with carbohydrates of prebiotic importance. Microb Ecol Health D. 2000;12:27-34.

38. Valeur N, Engel P, Carbajal N, Connolly E, Ladefoged K. Colonization and immunomodulation by *Lactobacillus reuteri* ATCC 55730 in the human gastrointestinal tract. Appl Environ Microb. 2004;70:1176–1181.

39. Alwazeer D, Cachon R, Divies C. Behavior of *Lactobacillus plantarum* and *Saccharomyces cerevisiae* in fresh and thermally processed orange juice. J Food Protect. 2002;65:1586–1589.

40. Huang Y, Adams MC. *In vitro* assessment of the upper gastrointestinal tolerance of potential probiotic dairy propionibacteria. Intl J Food Microbiol. 2004;91:253–260.

41. Leverrier P, Fremont Y, Rouault A, Boyaval P, Jan G. *In vitro* tolerance to digestive stresses of propionibacteria: Influence of food matrices. Food Microbiol. 2005;22:18–11.

42. Lorca GL, Font de Valdez G. A low-pH-inductible, stationary-phase acid tolerance response in *Lactobacillus acidophilus* CRL 639. Curr Microbiol. 2001;42:21–25

43. Hartke A, Bouch S, Giard JC, Benachour A, Boutibonnes P, Auffray Y. The lactic acid stress response of *Lactococcus lactis* subsp. lactis. Curr Microbiol. 1996;33:194–199.

44. Van de Guchte M, Serror P, Chervaux C, Smokvina T, Ehrlich S, Maguin E. Stress responses in lactic acid bacteria. A van Leeuw. 2002;82:187–216.

45. Champagne CP, Gardner NJ. Effect of storage in a fruit drink on subsequent survival of probiotic *lactobacilli* to gastro-intestinal stresses. Food Research International. 2008;41:539–543.

46. Li Y, Gue Ch, Yang J, Wei J, Xu J, Cheng Sh. Evaluation of antioxidant properties of pomegranate peel extract in comparison with pomegranate pulp extract. Food Chem. 2006;96:254-260.

47. Corcoran BM, Stanton C, Fitzgerald GF, Ross RP. Survival of probiotic lactobacilli in acidic environments is enhanced in the presence of metabolizable sugars. Appl Environ Microb. 2005;71:3060–3067.

48. Mainville I, Arcand Y, Farnworth E. A dynamic model that simulates the upper gastrointestinal tract for the study of probiotics. Int J Food Microbiol. 2005;99:287–296.

49. Khosravi-Darani K, Vasheghani-Farahani E, Shojaosadati SA. Application of the Plackett-Burman Design for the Optimization of Poly(β-hydroxybutyrate) Production by *Ralstonia eutropha*. Iran J Biotechnol. 2003;1(3):155-161.

50. Khosravi-Darani K, Vasheghani-Farahani E, Shojaosadati SA. Application of the Taguchi Design for Production of Poly(β-hydroxybutyrate) by *Ralstonia eutropha*, Iran J Chem Chem Eng. 2004;23(2):131-136.

3

Production and Partial Characterization of Pectinase from Snail (*Archachatina marginata*) Gut Isolates

E. C. Egwim[1*] and T. Caleb[1]

[1]*Biochemistry Department, Global Institute for Bioexploration, Federal University of Technology, P.M.B. 65, Minna, Niger state, Nigeria.*

Authors' contributions

This work was carried out in collaboration between both authors. Author ECE designed the study, wrote the protocol and supervised the study. Author TC managed the analyses of the study, the literature searches, statistical analysis and wrote the first draft of the manuscript. Both authors read and approved the final manuscript.

Editor(s):
(1) Anil Kumar, School of Biotechnology Devi Ahilya University, Madhya Pradesh, India.
(2) Kuo-Kau Lee, Department of Aquaculture, National Taiwan Ocean University, Taiwan.
Reviewers:
(1) Anonymous, Nigeria.
(2) Anonymous, Brazil.
(3) Anonymous, South Africa.
(4) Anonymous, Slovakia.

ABSTRACT

Aim: To identify isolates for industrial production of pectinase.
Methodology: The isolates were screened for pectinolytic activity using pectin as substrate. Enzyme activity was expressed as mg of glucose equivalent released per ml of crude enzyme solution per second. The kinetic parameters of the enzyme were determined to obtain the optimum pH, temperature, K_m and V_{max}. Pectinase produced from *Bacillus subtillis* was immobilized on chitosan and its activity was compared with that of free enzyme.
Results: Pectinase from *Bacillus subtilis* was screened to have the highest enzymatic activity among bacterial isolates while pectinase from *Aspergillus niger* had the highest enzymatic activity among fungal isolates. High yield of pectinase enzyme was obtained from *B. subtilis* after 24hrs with activity of 4.01×10^{-4} mg/ml/sec while high yield of pectinase was obtained from *A. niger* on the 5th day with activity of 2.07×10^{-4} mg/ml/sec. The optimum pH for pectinase produced from *B. subtilis* and *A. niger* were 8 and 6, respectively. The optimum temperature for pectinase from *B. subtilis* and *A. niger* were at 50ºC and 40ºC, respectively. The V_{max} and K_m of pectinase from *B. subtilis* and

---

Corresponding author: Email: c.egwim@futminna.edu.ng, evanschidi@gmail.com

A. niger were 16.88×10^{-4} (mg/ml/sec) and 10 (mg/ml); 9.29×10^{-4} (mg/ml/sec) and 30 (mg/ml), respectively. Optimum temperature and pH of immobilized pectinase were 70°C and 4.0, respectively. Residual activity of immobilized enzyme was 92% after storage at 4°C for 14 days. **Conclusion:** This study revealed that *Bacillus subtilis* from snail gut may be considered as a good candidate for industrial production of pectinase.

Keywords: Snail gut isolates; pectinase; enzyme activity; Km and Vmax; immobilized pectinase.

1. INTRODUCTION

Terrestrial gastropods feed on fresh plants with high protein, carbohydrate and calcium content [1] and may also participate with other soil invertebrate to decompose leaf litters [2]. Consequently, they need a large set of polysaccharide depolymerase such as pectinases for digestion of plant materials. The dependence of snail on microbial activity within their gut would explain their extra ordinary efficiency in digestion of up to 60-80% plant fiber [3]. Pectic substances are complex high molecular mass glycosidic macromolecules found in higher plants. They are present in the primary cell wall and are the major components of the middle lamellae, a thin extracellular adhesive layer formed between the walls of adjacent young cells. They are largely responsible for the structural integrity and cohesion of plant tissues [4]. Three major pectic polyssacharide groups are referred to as, homogalacturonan (HG) [5], rhamnogalacturonan I (RGI) [6], rhamnogalacturonan II (RGII) [7], all containing D-galacturonic acid to a greater or a lesser extent. Each of these pectic polyssacharides is digested by a particular pectinase enzyme which could be one of the following; protopectinase, pectin methyl esterase [8], pectin acetyl esterase, polymethylgalacturonase, polygalacturonase (PG), polygalacturonan acetyl esterase, pectatelyase and xylogalacturonan hydrolase [9].

Pectinases constitute a unique group of enzymes which catalyze the degradation of pecticpolymers present in plant cell wall [10]. It digests a galacturonic acid polymer by breaking the α-1,4glycosidic linkages between the giant molecules [11]. Through this process, it softens the cell wall and increase the yield of juice extract from the fruits. The major sources of enzyme include; plant and microorganism. From both technical and economical point of view, microbial sources of pectinase have become increasingly important. Of the hundred or so enzymes being used industrially, over a half is from fungi and yeast and over a third is from

bacteria with the remainder divided between animal (8%) and plant (4%) sources [12]. Pectinolytic enzymes have been reported to be produced by a large number of bacteria and fungi such as *Bacillus subtilis*, *Clostridium* spp., *Pseudomonas* spp., *Aspergillus* spp., *Penicillium* spp., *Thermomyces lanuginosus*.

Pectinase has a tremendous industrial applications especially, in food industry, for extraction and clarification of both sparkling clear juice (apple, pear, grapes and wine) and cloudy (lemon, orange, pineapple and mango) juice and maceration of plant tissue [13,14].

Despite the usefulness of pectinase, the enzyme is not produced at commercial quantities in Nigeria. However, snails are natural degrader of plants, it is expected that microorganisms inhabiting their gut should be efficient in hydrolysing plants component like pectin. Therefore, the present study was undertaken to identify snail gut isolates which may become good candidates for industrial production of pectinase.

2. MATERIALS AND METHODS

2.1 Collection of Samples

Five snails (5) were obtained from Kure Ultra-Modern market in Minna, Niger State, Nigeria. They were placed in a clean, covered bowl, containing sand, some vegetables and water. They were kept safe in the laboratory, prior to use.

2.2 Isolation of Microorganisms from the Gut of Snail

The snails were washed with tap water and outer shell rinsed with 96% ethanol for sterilization. The shells were broken (aseptically, by using flame and hand gloves) to remove the flesh and then dissected to reveal the gut. A portion of the gut was streaked onto nutrient agar plates, using inoculating loop and incubated for 18-24 hours at 37°C for bacterial growth, another portion was streaked in Saubourand Dextrose Agar (SDA) plates and incubated at 25°C for 3-5days for fungal growth. Individual colonies observed were sub-cultured on Nutrient Agar and Saubourand Dextrose Agar plates for another 18-24 hours and 3-5 days respectively and finally grown on agar slants to preserve the pure cultures. Identification of bacterial isolates was based on morphological (growth pattern) and biochemical test (Gram staining, catalase and urease test). The suspected organisms were then grown on selective media. Different fungi colonies were identified morphologically and by its microbiological appearance including the pigmentation, shape, presence of special structures and characteristics of spores. A small portion of each mycelia growth was carefully picked with the aid of a sterile needle and placed on a drop of lactophenol cotton blue on a slide and covered with cover slip. The slide was then observed under the microscope with magnification of x 10 and x 40 to detect the spores and some special structures of the fungi. This was followed by catalase test and Gram's staining technique to identify the bacterial isolated [15].

2.3 Screening of Isolates for Pectinase Activity

The isolates were screened for pectinase activity. One gram (1 g) of pectin was added in to SDA and Nutrient agar. These media were sterilized and distributed aseptically in petri dishes and the isolates were inoculated on the plates. The plates were observed for a zone of clearance after 3-5 days for fungi and 18-24 hours for bacteria. Isolates showing clear zone were considered as pectinase producers [16].

2.4 Quantification of Pectinolytic Potential of the Isolates

The isolates that showed zone of inhibition on plate were subculture in a nutrient broth to quantify the isolate potential for pectinase production. They were incubated at 37°C for 24 hours for bacteria isolates and for 5days at room temperature for fungi isolates. After the stipulated period of growth, the pectinolytic activity was determined by measuring the absorbance of reducing sugar using dinitrosalicylic acid (DNSA) reagent [17]. This was done by centrifuging the culture broth at 8,000 RPM for 5 and 15 minutes for bacteria and fungi, respectively. The supernatant was filtered using Whatman No. 1 filter paper and Pectinase enzyme was assayed by measuring half milliliter of 1% pectin in 0.1M citrate buffer (pH 5.8) in a test tube and 1.0 ml of culture filtrate was added. The reaction mixture was incubated at 50°C for 30 minutes and the reaction terminated by adding 3ml DNSA reagent. The tubes were heated in boiling water bath for 15 minutes and then cooled to room temperature. Then, the absorbance was read at 540 nm. The absorbance of the reducing sugar was extrapolated from glucose standard curve and the enzyme activity was expressed as mg of glucose equivalent released per ml/sec [17].

2.5 Pectinase Production Profile from Growing Culture of Snail Gut Isolates

The isolates showing pectinase activity were placed in a basal medium, the medium consist of $NaNO_3$, 2%; K_2HPO4, 1%; $MgSO4$, 5%; KCl 5%; $FeSO4$, 0.001%; Pectin 15%. The culture was grown for 7days at 25°C for fungi and 30 hours for bacteria. The culture broth was sampled and assayed every 24 hours for fungi and 6 hours for bacteria.

2.6 Microbial Biomass Determination

The bacterial growth was determined by spectrophotometric method. This was carried out by growing the isolates in nutrient broth and sampled to determine the optical density (OD) at 575 nm; it was carried out simultaneously with enzyme production profile. This measurement of turbidity or optical density is an indirect measurement of cell biomass that includes both living and dead cell [18].

2.7 Effect of Temperature on Pectinase Activity

The optimum temperature for pectinase was determined by incubating 1 ml of the crude enzyme with 2 ml 3% pectin in citrate buffer, pH (5.8) at different temperature (30-90°C) for 30 minutes. Reducing sugars were estimated by the dinitrosalcyclic acid reagent method [19].

2.8 Effect of pH on Pectinase Activity

The optimum pH was determined by incubating 1 ml crude enzyme with 2 ml of 3% pectin in 1 ml citrate buffer at different pH (3-8) for 30 minutes at 50°C. Reducing sugars thus released were estimated spectrophotometrically by dinitrosalicylic acid reagent method.

2.9 Effect of Substrate Concentration on Pectinase Activity

Crude enzyme (1ml) was added to 2ml of different substrate concentrations, 1%, 1.5%, 2.0%, 2.5%, 3.0%, 3.5% (W/V) and incubated at appropriate temperature and pH obtained above. The reducing sugar was estimated spectrophotometrically, Vmax and Km were then determined from Line Weaver-Burk plot.

2.10 Immobilization of Pectinase on Chitosan

Pectinase produced from *Bacillus subtilis* was immobilized by physical adsorption on Chitosan bead as described by [20]. Chitosan powder (2 g) was soaked in hexane under agitation conditions (100 rpm) for 1 hour. Excess hexane was removed by decantation followed by the addition of 10ml of pectinase crude enzyme. The pectinase was on the support under agitation for 3 hours at room temperature followed by additional period of 18 hours under static conditions at 4°C. The derivative was filtered (Whatman filter paper 41) and thoroughly rinsed with hexane. The optimum enzyme loading capacity of the chitosan and the loading time were determined. The effects of other factors, such as optimum pH, temperature and enzyme storage stability of immobilized enzyme activity were investigated as described in reference [20]. Optimum pectinase loading time was determined by sampling the bead-enzyme mixture at an hour time interval for 6 hours. The enzyme activity was assayed after centrifuging the sample at 1500 rpm for 30 minutes to remove the carrier. Storage stability of immobilized pectinase was determined from its decay constant and half-life [21].

3. RESULTS AND DISCUSSION

The bacteria isolated from the gut of snail include *B. subtilis, streptococcus sp,* and *S. aureus* while fungal isolates were *A. niger, P. notatum, Trychophyton violacium* and *Torulopsis datilla.*

3.1 Screening and Quantification of Pectinolytic Potential of the Isolates from Snail Gut

All the bacteria and only two of the fungal isolates showed zone of clearance on pectin agar plates, thus indicating the ability of these organisms to produce pectinase. From the bacteria screened for pectinase production, *Bacillus subtilis* showed the highest enzymatic activity (3.0×10^{-4} mg/ml/sec). The pectinase activity of *A. niger* (2.6×10^{-4} mg/ml/sec) was greater than that of *P. notatum* after secondary screening as shown in Table 1.

3.2 Biomass Yield-enzyme Activity Profile

Bacillus subtilis gave the highest enzyme yield after 24 hours as shown in Fig. 1, with an enzyme activity of 4.01×10^{-4} mg/ml/sec. *A. niger* gave the highest enzyme yield after 5 days with activity of 2.06×10^{-4} mg/ml/sec as shown in Fig. 2. *Bacillus subtilis* showed its highest enzymatic activity (4.01×10^{-4} mg/ml/sec) after 24 hours with highest biomass growth after 20 hours (Fig. 1). This is in correlation to findings of Kashyap et al. [22] who observed pectinolytic activity after 24 hours for *Bacillus subtilis* but the result is contrary to Abdul et al. [23] who observed maximum pectinase activity from *B. subtilis* at 48 hours with the activity of 1700 µ/ml. It was observed that, the time required by the pectinase from snail gut isolate (*B. subtilis*) to develop peak activity was significantly shorter than the time reported by Abdul et al. [23]. Therefore, the Bacillus isolate in the present study may be considered a better source of pectinase production. The difference observed in the present result could be due to difference in strain of *B. subtilis* used. Pectinase activity increased, with exponential growth of the isolates, the decrease with time of pectinase activity during incubation was regarded as the death phase of the isolates. *Aspergillus niger* exhibited its highest enzymatic activity on the 5th day of incubation with an activity of 2.07×10^{-4} mg/ml/sec and then declined to the 7th day of incubation. A similar result was reported by Anithara [24] who recorded maximum pectinase activity after 4 days. The slight difference observed could be due to the difference in substrate and or difference in strains of *A. niger.* Therefore, the incubation period at which pectinase had maximum activity (24 hours for *Bacillus subtilis* and 5 days for *A. niger*) is the best time at which pectinase should be harvested from these isolates which is the period of maximum production.

3.3 Effect of Temperature on Pectinase Activity

The optimum temperature of pectinase from *B. subtilis, A. niger, P. notatum* were 50°C, 40°C and 60°C respectively, with activities of 0.43 ×

10^{-4}, 0.44 × 10^{-4} and 0.27 × 10^{-4} mg/ml/sec respectively as shown in Fig. 3. Optimum temperature of pectinase from *Bacillus subtilis* observed in the present study agreed with the findings of Eleni et al. [25] who recorded optimum temperature between 45°C and 55°C for pectinase from *B. subtilis*. Optimum temperature attained by pectinase produced from snail gut *Aspergillus niger* supports the findings of Mohammed et al. [26] who reported optimum temperature of 40°C. Pectinase from *Penicillium notatum* showed highest activity at optimum temperature of 60°C which agreed with the finding of Sinitsyna et al. [27]. Increased in temperature resulted to an increase in pectinase activity up to a certain point (optimum temperature), where activities decreased as the temperature increased further. Every enzyme has optimum or apparent temperature at which its activity is highest, above this range, the enzyme is denatured, therefore losses its active site for catalysis. Below this range, the enzyme is less active due to intra-molecular hydrogen bond within the enzyme for catalysis to take place.

3.4 Effect of pH on Pectinase Activity

The optimum pH of pectinase from *B. subtilis, A. niger, P. notatum* were 8, 6 and 4, respectively, with activities of 2.56×10^{-5}, 1.30×10^{-4} and 1.17×10^{-4} mg/ml/sec, respectively (Fig. 4). Pectinase from *Bacillus subtilis* has its optimum activity at pH that correlates to the finding of Kashyap et al. [22] but contrary to Soriano et al. [29] who reported optimum pH at 10.0 for pectinase from *Bacillus subtilis*. Optimum pH (6) for pectinase from *A. niger* (with activity of 1.3 x 10^{-4} mg/ml/sec) was in agreement to the finding of Oyeleke et al. [28] who reported optimum pH at 6.0 for pectinase from *A. niger*.

Optimum pH of pectinase from *Penicillium notatum* agreed with Alana et al. [29] and Sinitsyna et al. [27]. The pH above or below this range alter the conformation of the enzyme and ultimately its activity, therefore compromising the binding of the enzyme with substrate.

Table 1. Pectinase activity of the snail gut isolates

Isolates	Enzyme activity (mg/ml/sec) × 10^{-5}
S. aureus	0.10
B. subtilis	3.00
Streptococcus spp	1.10
Aspergillus niger	2.60
Penicillium notatum	2.20

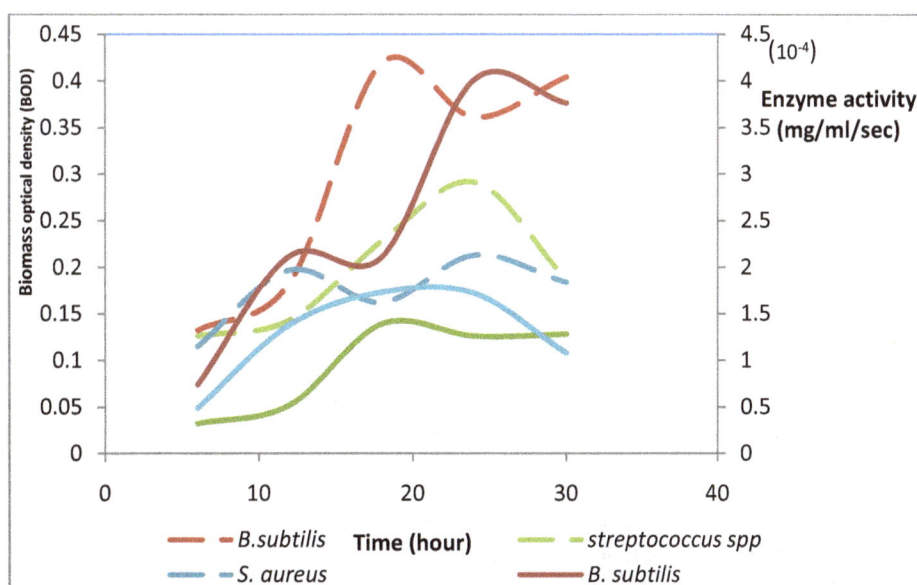

Fig. 1. Pectinase activity (mg/ml/sec) and biomass optical density of *B. subtilis, streptococcus spand S. aureus*

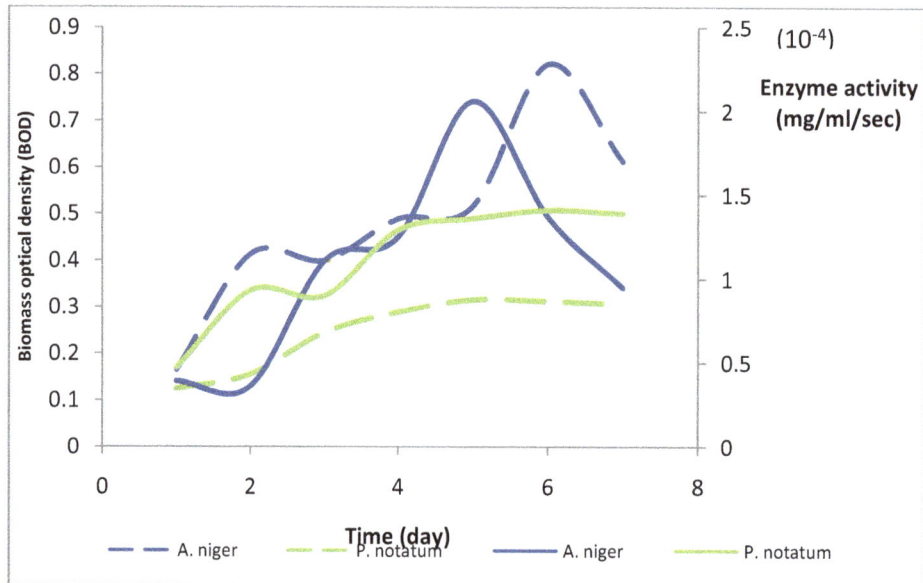

Fig. 2. Pectinase activity (mg/ml/sec) and biomass optical density of *A. niger* and *P. notatum*

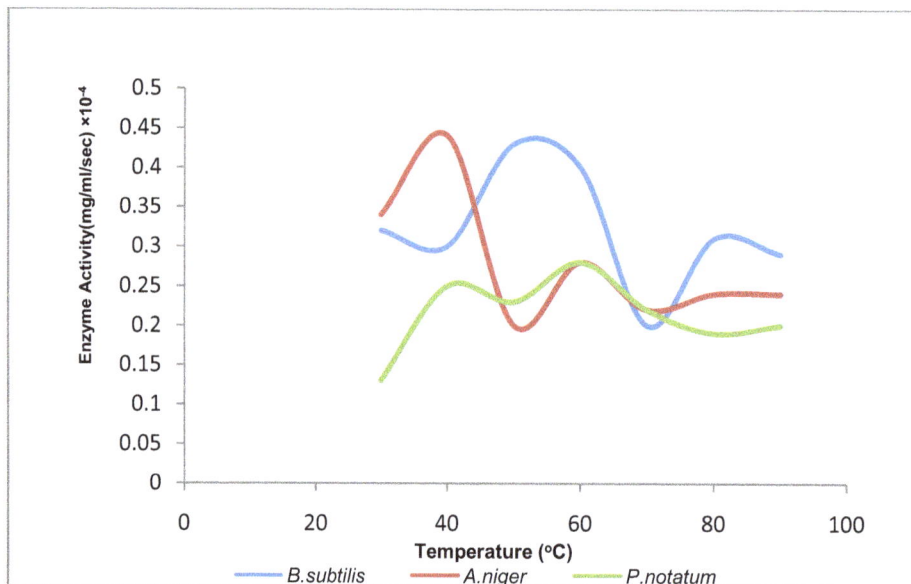

Fig. 3. Temperature profile of pectinase enzymes from *B. subtilis*, *A. niger* and *P. notatum* isolated from snail gut.

3.5 Effect of Substrate Concentration on Enzyme Activity

The K_m and V_{max} values of pectinase from different sources were calculated using Line weaver-Burk plot and were shown in Table 2. This result indicated that, pectinase from *Bacillus subtilis* has the highest affinity for substrate due to its lowest K_m, it also has the highest utility of pectin substrate as a result of its highest V_{max}.

The present finding is similar to the finding of Anna et al. [30] who reported that pectinase from *Bacillus firmus* was better than that isolated from *Aspergillus niger*. Due to this high binding of pectinase from *B. subtilis* with pectin substrate, small quantity of the enzyme will digest a considerably high amount of substrate. This may therefore reduce the cost for the enzyme in industrial use.

Furthermore, the present results were compared with the observation of Adesina et al. [31] who reported the V_{max} of pectinase produced by *Rhodotorulla* spp and *Mucor mucorales* to be 0.023304 µ/mg/min and 0.043304 µ/mg/min, respectively. The maximum velocities (V_{max}) of pectinase from *A. niger* and *B. subtilis* isolated from the gut of snail were higher than the value recorded by Adesina et al. [31]. This suggested that pectinase produced from snail gut isolates has better utility of its substrate.

3.6 Characterization of Immobilized Pectinase

3.6.1 Effect of pectinase loading time

The effect of loading time on Chitosan support is shown in Fig. 5. The maximum enzyme loading time was obtained at 4 hours with enzyme activity of 1.18×10^{-4} (mg/ml/sec).At this loading time, the intermolecular spaces of the chitosan (carrier) have been completely adsorbed with the

enzyme and beyond this time, the activity of immobilized pectinase remained constant.

3.6.2 Effect of temperature on immobilized pectinase

The optimum temperature of immobilized pectinase from *B. subtilis* was 70ºC while optimum temperature of free pectinase from *B. subtilis* was 50ºC. This result supported the findings of Egwim et al. [20] who showed that the optimum temperature of immobilized lipase increased from 35ºC for free enzyme to 40ºC. The result also supported Flashmy et al. [32] who reported optimum temperature of immobilized urease increased from 55ºC for free enzyme to 65ºC. The increase in temperature in the immobilized enzyme over free enzyme in the present study suggested that immobilized enzyme has a higher resistance to changes in temperature than free enzyme. The stability of the enzyme up to 70ºC could make it less susceptible to thermal inactivation during industrial processes.

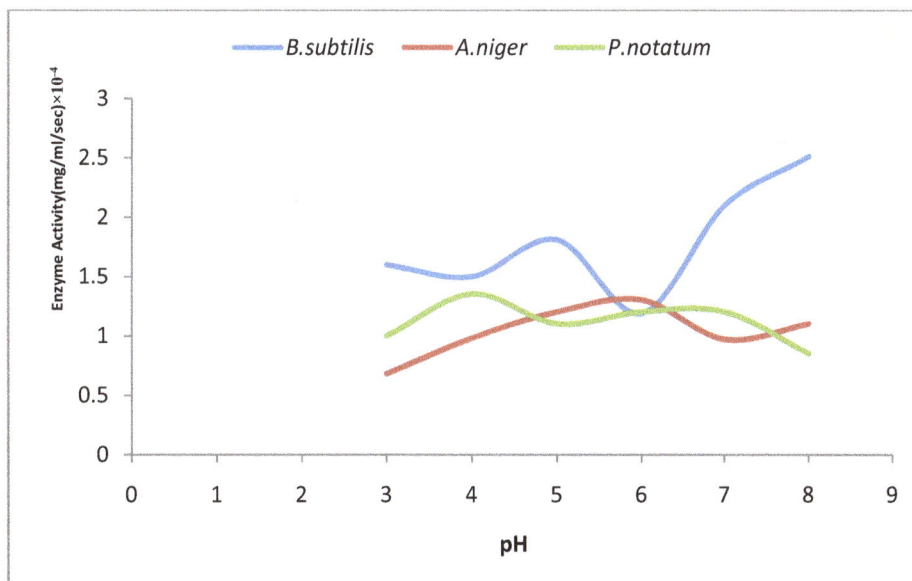

Fig. 4. pH profile of pectinase enzymes from *B. subtilis*, *A. niger* and *P. notatum* isolated from snail gut

Table 2. V_{max} and K_m of pectinases from pectinolytic isolates

Isolates	V_{max}(mg/ml/sec) ×10⁻⁵	K_m (mg/ml)
B. subtilis	16.88	10
S. aureus	0.88	410
Streptococcus sp	3.62	60
Aspergillus niger	9.29	30
P. notatum	1.10	630

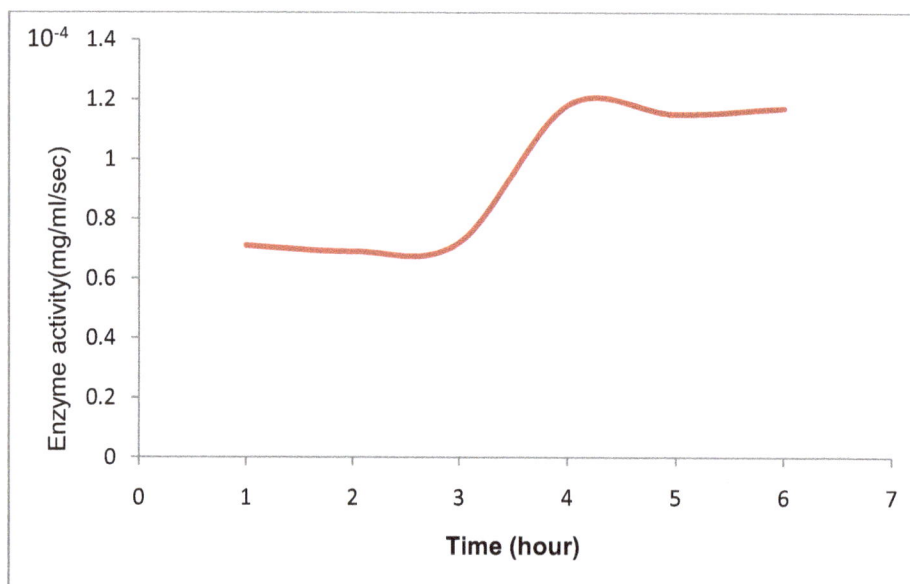

Fig. 5. Pectinase loading capacity and loading time on chitosan

3.6.3 Effect of pH on immobilized pectinase

The optimum pH of immobilized pectinase was found to be at 4.0. Optimum pH of immobilized pectinase was shifted to acidic range (pH 4.0) from pH 8.0, which was the optimum pH for free enzyme. This acidic shift in the optimum pH was in agreement with the general observation that the positively-charged supports displace pH-activity curve of the enzyme attached to them towards lower pH values [33]. The stability of pH in acidic region could be an advantage in fruit processing such as citrus fruits as a result of their acidic nature.

3.6.4 Storage stability of immobilized enzyme

The residual activity of immobilized pectinase achieved after storage for 14 days was 92%. Thus, it would take 132 days for immobilized pectinase to attain half of its activity (half-life). These results are similar to the findings of Egwim et al. [20] who reported 96% residual activity of immobilized lipase after storage for 90 days. Furthermore, the storage stability observed was higher than the previous study of Li et al. [21] who reported that half-life of immobilized pectinase was 30 days under 4°C condition. The higher half-life of immobilized pectinase produced by *B. subtilis* isolated from snail gut could be found advantageous in industrial applications because, most enzymes are lost after first use while other loss activity after a few days of storage. This difference could be as a

result of the effect of support use for immobilization on enzyme storage or the nature of the enzyme itself. The half-life indicated possible reusability of the immobilized enzyme while residual activity showed the potency of the immobilized enzyme after each cycle. The storage stability of immobilized enzyme could be due to intramolecular linkage conferred by the carriers (chitosan) and depends on the length of such linkage [34]. It could also be due to the restriction of the enzyme within the carrier. Therefore, for industrial application, pectinase immobilized on chitosan bead can be reused over a long period of time and can withstand harsh industrial conditions (high temperature and acidic medium).

3.6.5 Decay constant (κ) and half -life (T1/2)

The κ and T½ values for immobilized pectinase enzyme were computed as follow;

$$de/dt = -KE \qquad (1)$$

$$At\ t=0;\ E=E_o$$

Integrating

$$E = E_o\ e^{(-Kt)} \qquad (2)$$

Where E_o is the initial active enzyme concentration and t is time elapsed during reaction. The residual enzyme activity A_r is directly proportional to the concentration of the active enzyme (E):

$$A_r A_o = E/E_o \qquad (3)$$

From equation 2&3, the residual enzyme activity follows as:

$$Ar = Aoe^{(-Kt)} \qquad (4)$$

From equation 4;

$$K = 2.303/t \log A_r/A_o \qquad (5)$$

Enzyme half-life (t1/2) is the time lapse for the enzyme activity to decrease by 50% of the initial value. From equation (5) it means that half life (t1/2) can be calculated.

$$T1/2 = \ln (0.5)/K = 0.693/K \qquad (6)$$

$A_o = 0.92 \times 10^{-4} (mg/ml/sec)$
$_r = 0.99 \times 10^{-4} (mg/ml/sec)$
t = 14days
Using equation 5;
$K = 2.303/t \times \log(A_o/A_r)$
$K = -0.00524$
Therefore, half-life can be calculated using equation 6
$T_{1/2} = \ln (0.5)/K$
$T_{1/2} = 132$ days.

4. CONCLUSION

It can be concluded that *Bacillus subtilis* and *Aspergillus niger* obtained from snail gut may be good candidates for industrial production of pectinase.

ACKNOWLEDGEMENTS

Our sincere gratitude to the Department of Microbiology and Department of Biochemistry of Federal University of Technology, Minna for providing the chemicals and materials used during this research.

REFERENCES

1. Chevalier L, Coz-bouhnik M, Charrier M. Influence of inorganic compounds on food selection by the brown garden snail (*Gastropoda: Pulmonata*). Malacologia. 2002;45(2):125-32.

2. Hatziioannou M, Eleutheriadis N, Lazaridou-Dimitriadou M. Food preference and dietary overlap by terrestrial snail in Lagos Area. J. Mollusc Study 1994;60:331-341.

3. Charrier MY, Gerard F, Gaillard-Martinie B, Kader A, Gerard A. Isolation and characterization of cultivable fermentative bacteria from the intestine of two edible snails, *Helix pomatia* and *Cornu aspersum* (*Gastropoda: Pulmonata*). Biol. Res. 2006;97:16-60.

4. Alkorta I, Gabirsu C, Lhama MJ, Serra JL. Industrial applications of pectic enzymes. A review. Process Biochem. 1998;339:21-8.

5. Jayani RS, Saxena S, Gupta R. Microbial pectinolytic enzymes. Process Biochem. 2005;40(4):2931-44.

6. Willats WGT, Knox P, Mikkelsen JD. Pectin: new insights into an old polymer are starting to gel. Trends Food Sci. Technol. 2006;17:97-104.

7. Vincken JP, Schols HA, Oomen RJFJ. If homogalacturonan were a side chain of rhamnogalacturonan I: Implications for cell wall architecture. Plant Physiol. 2003;132:1781-9.

8. Coutinho PM, Henrissat B, Gilbert HJ, Davies G, Henrissat B, Svensson B. Carbohydrate-active enzymes: An integrated database approach. In Recent Advances in Carbohydrate Bioengineering. The Royal Society of Chem. 1999;3-12.

9. Vlugt-Bergmans CJB, Meeuwsen PJA, Voragen AGJ, Van Ooyen AJ. Endo-xylogalacturonan hydrolase, a novel pectinolytic enzyme. Appl. Environ. Microbiol. 2000;66(1):36-41.

10. Oyeleke SB, Oyewole OA, Egwim EC, Dauda BEN, Ibeh EN. Cellulase and pectinase production potential of *Aspergillus niger* isolated from corn cob. Bayero J of Pure and Appl. Sci. 2012;5(1):78-83.

11. Oyeleke SB, Egwim E C, Auta SH. Screening of *Aspergillus flavus* and *Aspergillus fumigates* strains for extracellular protease enzyme production. J. Microbiol.and Antimicrobials. 2010;2(7):83-7.

12. Chaplin, M. Sources of enzymes. Faculty of Engineering, Science and the Built Environment. London South Bank University. 2004;35(2):91-6.

13. Kashyap DR, Vohra PK, Tewari R. Application of pectinases in the commercial sector: A review. Bioresour. Technol. 2001;77:215-27.

14. Hoondal GS, Tiwari RP, Tewari R, Dahiya N, Beg QK. Microbial alkaline pectinases

and their industrial applications: A review. Appl. Microbiol. Biotechnol. 2002;59:409-18.

15. Oyeleke SB, Manga BS. Essentials of Laboratory Practicals in Microbiology First Edition, Tobest Publisher, Minna, Niger State, Nigeria. 2008;15-34.

16. Palaniyappan M, Viljayagopal VJ, Viswanathan R, Viruthagiri T. Screening of natural substrate and optimization of operating variables on the production of pectinase by submerge fermentation using *Aspergillus niger* MTCC 281, Afri. J Biotechn. 2009;8(4):682-6.

17. Bertrand TF, Fredric T, Robert N. Production and partial characterization of a thermostable amylase from Ascomycetes yeast strain isolated from starchy soil. MC Graw-Hill Inc., New York. 2004;12(1):53-5.

18. Koch AL. Turbidity measurement of bacterial in some available commercial instruments. Anal. Biochem. 1990;38(3): 252-9.

19. Miller GL. Use of dinitrosalicylic acid reagent for determination of reducing sugar. Anal. Chem. 1959;(31):426-8.

20. Egwim EC, Adesina M, Austin A, Oyewole OA, Okoliegbe IN. Optimization of Lipase Immobilized on Chitosan Beads for Biodiesel Production. Global Research J. Microbiol. 2012;(2):103–12.

21. Li Xiu-jin, ZhongFei, Xiao Yue-juan, Guo Hong-yan. Immobilization of pectinase on chitosan. Food Sci. 2002;23(10):50-3.

22. Kashyap DR, Chandra S, Kaul A, Tewari R. Production, purification and characterization of pectinase from a *Bacillus* sp. DT7. World J Microbiol. Biotechnol. 2000;16:277-82.

23. Abdul SQ, Muhammed A, Bhutto YC, Imrana KM, Umar D, Safia SB. Production of pectinase by *Bacillus subtilis* EFRL 01 in a date syrup medium. Afri. J. Biotech. 2012;11(62):12563-70.

24. Anithara R, Mrudula S. Pectinase production in solid state fermentation by *Aspergillus niger* using orange peel as substrate. Glob. J. Biotech. Biochem. 2011;6(2):64-71.

25. Eleni G, Marcia MCN, Soares RS. Screening of bacterial strains for pectinolytic activity; characterization of the polygalacturonase produced by *Bacillus sp*. Revista de Microbiologia. 1999;30:29-30.

26. Mohammed L, Buga SI, Andrew SN. Partial polygalacturonase from *A. niger*. Afri. J. Biotech. 2010;9(52):8944-54.

27. Sinitsyna OA, Fedorova EA, Semenova MV. Isolation and characterization of extracellular pectin lyase from *Penicillium canescens*. Biochem (Moscow). 2007;72(5):565-71.

28. Soriano M, Diaz P, Pastor FIJ. Pectinolytic systems of two aerobic sporogenous bacterial strains with high activity on pectin. Curr. Microbiol. 2006;50:114-8.

29. Alana A, Alkorta I, Dorminguez JB, Llama MJ, Serra JL. Pectin lyase activity in *penicillium italicum* strain. Appl Environ. Microbiol. 1990;56:3755-9.

30. Roosdiana A, Prasetyawan A, Mahdi C. Sutrisno. Production and characterization of *Bacillus firmus* pectinase. J. Pure App. Chem. Res. 2013;2(1):35-41.

31. Adesina M, Felicia CI, Adefila O, Aewale AI, Adeyefal O, Ummi H, Shadrach O. Production of pectinase by fungi isolated from degrading fruits and vegetable. Nat Sci. 2013;11(10):102-8. ISBN: 1545-0740.

32. Fashmy AS, Bagos VB, Mohammed TM. Immobilization of *Citrullus vulgaris* urease on cyanuric chloride DEAE-cellulose ether, preparation and properties. Biores. Tech. 1998;64:121-9.

33. Krajewska B, Leszko M, Zaborska W. Urease immobilized on chitosan membrane: Preparation and properties. J. Chem. Tech. Biotech. 1990;48:337-50.

34. Torchilin VP, Malcsimenko AV, Marchiek VN. Principle of enzyme stabilization. The possibility of enzyme self-stabilization under the action of potential reversible intermolecular cross-linkage of different length. Biochem. Biophy. Acta. 1979;568(2):1-10.

4

Alkaline Cellulase Production by *Penicillium mallochii* LMB-HP37 Isolated from Soils of a Peruvian Rainforest

Karin Vega[1], Victor Sarmiento[1], Yvette Ludeña[1], Nadia Vera[1], Carmen Tamariz-Angeles[2], Gretty K. Villena[1] and Marcel Gutiérrez-Correa[1*]

[1]Laboratorio de Micologíay Biotecnología, Universidad Nacional Agraria La Molina, Av. La Molina s/n, La Molina, Lima 12, Peru.
[2]Laboratorio de Biología, Facultad de Ciencias, Universidad Nacional Santiago Antúnez de Mayolo, Huaraz, Peru.

Authors' contributions

This work was carried out in collaboration between all authors. Authors KV, VS and YL carried out fermentation experiments, enzyme determinations and performed the statistical analysis. Author NV collected the soil samples and helped in fungal DNA extraction. Author CT-A managed the molecular alignment and phylogenetic analysis. Authors MG-C and GKV conceived and designed the study and wrote the draft of the manuscript. All authors read and approved the final manuscript.

Editor(s):
(1) Chung-Jen Chiang, Department of Medical Laboratory Science and Biotechnology, China Medical University, Taiwan.
Reviewers:
(1) N. Murugalatha, Department of Applied Sciences and Humanities, Quantum School of Technology, Roorkee, Uttarakhand, India.
(2) Ali Mohamed Elshafei, Department of Microbial Chemistry, Division of Genetic Engineering and Biotechnology, National Research Centre, Egypt.

ABSTRACT

Alkaline cellulases are demanded by the textile industry for several purposes but commercial preparations showing activity at alkaline conditions are very scarce.
Aim: To characterize a Penicillium strain isolated form soils of a Peruvian rainforest showing alkaline cellulase activity that may be useful for the textile industry.
Methodology: The molecular identification was based on the DNA sequence of its ITS region using ITS1 and ITS4 primers after PCR amplification. Cellulase production was evaluated in shaken flasks by using either lactose or microcrystalline cellulose. Total cellulase (as FPA) and endoglucanase activities were evaluated by the standard methods at several pH levels. Also, the

*Corresponding author: mgclmb@lamolina.edu.pe

cellulase activity of culture filtrates was tested for antipilling activity as compared to a commercial neutral cellulase preparation.

Results: After raw data of ITS DNA sequence was processed, multiple alignment and phylogenetic analysis confirmed that our strain can be named as *Penicillium mallochii* LMB-HP37. Higher activity was attained for neutral total cellulase on lactose (3371±108 U/l at pH 7.4) and alkaline cellulases attained similar activity levels than the acid cellulase (2978±151 U/l at pH 8.4 and 2910±42 U/l at pH 9.4). FPA and endoglucanase activities were produced at high volumetric (46.8±1.5 and 13.5±1.0 U/l.h, respectively) and specific (32.9±1.1 and 9.5±0.7 U/gbiomass.h, respectively) productivities at the same pHs which indicate that this strain may be suitable for commercial development. The enzyme of *P. mallochii* LMB-HP37 had slightly better results than the commercial enzyme as an anti-pilling agent even though is a crude preparation.

Conclusion: *Penicillium mallochii* LMB-HP37 produced high total cellulase activity on lactose which compares to well-known cellulase producers but at neutral to alkaline pH levels. Data obtained reveal that the crude enzyme is suitable for anti-pilling process (biopolishing) and may be also useful for biostoning.

Keywords: Cellulase; endoglucanase; biopolishing; biostoning; fungi; Penicillium mallochii.

1. INTRODUCTION

Enzymes are fundamental components of all living things that play important roles in the metabolic reactions. Numerous industrial processes use enzymes as organic catalysts to speed up either conversion or synthesis reactions at lower costs and with reduced environmental problems [1]. For industrial scale applications, microbial enzymes from different sources are considered superior enzymes and nowadays over 150 industrial processes use either enzymes or microbial cells as catalysts. The total market for industrial enzymes reached $3.3 billion in 2010 and it is estimated to reach a value of 4.4 billion by 2015 [2]. Many enzymes are sold as bulk enzymes for the detergent, textile, pulp and paper industries, and in the biofuels industry, among others.

In the textile industry, enzymes are also increasingly being used to develop cleaner processes and reduce the use of raw materials and production of waste. The application of cellulases for denim finishing and laccases for decolorization of textile effluents and textile bleaching are the most recent commercial advances [3]. The textile industry requires highly stable enzymes, and good performance at extreme values of pH and temperature. New strategies are continuously emerging for the immobilization of enzymes with superior efficiency and usage [4]. The principal enzymes applied in textile industry are hydrolases and oxidoreductases. The group of hydrolases includes amylases, cellulases, proteases, pectinases and lipases / esterases [3,5,6].

Cellulases are well established in textile wet processing as agents for fiber surface modification [5]. Cellulases are multi enzyme systems produced by bacteria and fungi mostly present in soil environments. The widely accepted mechanism for enzymatic cellulose hydrolysis involves the synergistic activity of endoglucanase, exoglucanase or cellobiohydrolase, and β-glucosidase [7,8,9,10]. Cellulases are used in several textile processing activities like mercerization, scouring, bio-polishing, washing, stone finishing, pre-treatments of bast fibers, carbonization process, and wool scouring [11]. The enzymatic cleaning is carried out at neutral pH [3] and, although there are pectinases that work optimally at this pH, most commercial cellulases have optimal activity at acidic pH which implies that sequential treatment was done by adjusting the pH fiber at each stage. Cellulases have recently been found with optimal activity at neutral to alkaline pH, and these are gradually being incorporated into fiber treatment processes, especially in the polishing and washing of denim processing because the alkaline or neutral pH reduces textile back-staining [12,13,14].

Neutral and alkaline cellulases have been identified and isolated from bacteria [12,15,16] and fungi [13,17], and the search for new microorganisms is constant. We have isolated several fungal strains from soils of a Peruvian rainforest that show alkaline cellulase activities [18]. Here, we have further investigated the alkaline cellulase production of one *Penicillium* strain as candidate for commercial use in the textile industry.

2. MATERIALS AND METHODS

2.1 Strain

Strain *Penicillium* sp. LMB-HP37, isolated from soils of the undisturbed Macuya Forest, near Pucallpa (Peru), located at latitude 8°27'28.55"S and longitude 74°53'44.33"W, were used in this study [18]. The strain was conserved in potato dextrose agar (PDA) slants at 7°C. Spores of the selected strain were washed from 5-day PDA plates with 10 ml of 0.1% (v/v) Tween 80 solution and this suspension (1 x 10^6 spores per ml) was used as inoculum.

2.2 Molecular Characterization of Fungal Strain

The selected strain was subjected to molecular characterization based on the sequences of the internal transcribed spacer (ITS). Fungal strain was grown for 48 h on liquid culture medium with glucose (10 g/l) as carbon source (see below) at 28°C and 175 rpm. Approximately, 100 mg of fresh biomass was ground in liquid nitrogen and the DNA was extracted following the procedure of Möller et al. [19]. Highly purified genomic DNA was used as template for PCR amplification of the ITS region by using ITS1 (5' - TCCGTAGGTGAACCTGCGG'-3') and ITS4 (5' - TCCTCCGCTTATTGATATGC'-3) primers [20]. PCR was performed using the following procedure: 1 cycle at 95°C for 5 min; 25 cycles at 95°C for 1 min, 55°C for 1 min, and 72°C for 1.5 min; and ended with 1 cycle at 72°C for 5 min. Amplification products were purified and sent them to Macrogen Inc. (Seul, South Korea, http://www.macrogen.com) for sequencing. The resulting DNA sequences were read and edited with Chromas Lite versión 2.01 (Technelysium Pty Ltd, 2007) y CAP3. Nucleotide sequences were compared by the BLASTN search with those from type material of the NCBI GenBank database (http://www.ncbi.nlm.nih.gov/). Phylogenetic analysis was carried out with MEGA5 using neighbor-joining method with 1000 bootstraps. The nucleotide sequences obtained have been deposited in the GenBank database (accession number KP979777).

2.3 Culture Medium and Cellulase Production

Culture medium for growth and cellulase production had the following composition per liter: 2 g KH_2PO_4; 1.4 g $(NH4)_2SO_4$; 0.3 g urea;

0.3 g $CaCl_2.2H_2O$; 0.3 g $MgSO_4.7H_2O$; 1 g peptone; 0.2% (v/v) Tween 80; 5 mg $FeSO_4.7H_2O$; 1.6 mg $MnSO_4.2H_2O$; 1.4 mg $ZnSO_4.7H_2O$; 2 mg $CoCl_2.6H_2O$ and either 10 g lactose or 10 g microcrystalline cellulose (20 μm). The initial pH was 5.5.

Each 250 ml flask containing 70 ml production medium was inoculated with a 3% (v/v) of the above spore suspension. All inoculated flasks were incubated at 28°C in a shaker bath at 175 rpm. Four replicates were taken in all cases. Cultures were centrifuged at 3,000 rpm for 20 min; solids were used for biomass determination and supernatants were used for enzyme activity and soluble protein determinations.

2.4 Biomass Determination

For lactose cultures biomass was determined by measuring the dry cell weight. For cellulose cultures biomass was indirectly determined by measuring the intracellular protein contain as previously reported [18].

2.5 Enzyme Determinations

Total cellulase, measured as filter paper activity (FPA), and endoglucanase were determined as recommended by Ghose [21] by using different buffers (50 mM). For enzyme determinations at pH 4.8, 7.4 - 8.4, and 9.4 Na-acetate, HCl-barbital and glycine-NaOH buffers were used, respectively. An enzyme unit (U) is defined as the amount of enzyme that releases 1 μmol product (glucose equivalents) per min.

2.6 Protein Determination

Both soluble and intracellular protein were estimated by the Lowry et al. [22] procedure. For soluble protein determination culture supernatants were first precipitated with two volumes of 10% (w/v) trichloroacetic acid, vortexed, kept overnight at 4°C, and centrifuged at 6,000 rpm at 4°C for 25 min. Pellets were then resuspended in phosphate buffer saline.

2.7 Antipilling Assessment

The cellulase activity of culture filtrates were tested for antipilling activity. For this we used a commercial enzyme kindly provided by Química Nava S.A.C. (Lima, Peru) as reference enzyme since it works at pH 7. A 0.5 g piece of Pima cotton cloth was introduced into a specimen

Type 2 Container (SDL Atlas, U.S.A.) containing 50 stainless steel balls of 0.6 cm diameter (SDL Atlas), 13 endoglucanase units, non ionic detergent Idrosolvan RO 7 (Bozzetto Group, Italy) up to 50 ml and pH was adjusted to 7 with 10% Na_2CO_3. Containers were incubated at 60°C and 40 rpm for 45 min. Then cloths were withdrawn from the containers, thoroughly washed with deionized water and dried at 120°C for 3 min in a Type LTE lab dryer (Werner Mathis, Switzerland). The amount of fluffs was observed in a Model M227PAV Universal Pilling Assessment Viewer (SDL Atlas). Two replicates were run for each enzyme system and for control system without enzyme.

2.8 Statistical Analysis

Data were analyzed by Statgraphics Centurion XVI (Statpoint Technologies, Inc., Warrenton, VA, USA). Analysis of variance (ANOVA) by the General Linear Models procedure and Duncan's multiple range tests were used to find significant differences between treatments.

3. RESULTS AND DISCUSSION

The soils of the rainforest are home to a wide variety of bacteria and fungi that may have utility in many industrial processes. In the course of searching fungal strains with alkaline cellulase activity needed to textile industries, we isolated several strains from the soils of the Macuya rainforest (Pucallpa, Peru) [18]. Strain LMB-HP37 was preliminary reported as a high alkaline cellulase producer. Here we have molecularly identified this strain and its physiology related to cellulase production has been further studied.

Strain LMB-HP37 shows morphological characteristics similar to members of section

Sclerotiora of the *Penicillium* genus, i.e., monoverticillate conidiophores, greenish gray conidial masses and orange pigment in PDA plates (Fig. 1) [23]. The molecular identification was based on the DNA sequence of its ITS region using ITS1 and ITS4 primers. After raw data was processed, BLASTN analysis gave 99% identity with *Penicillium mallochii* DAOM 239917 (GenBank accession number NR_111674). Multiple alignment and phylogenetic analysis showed that our strain can be assigned to *Penicillium mallochii* (Fig. 2). *Penicillum levitum* CBS 345.48 (accession number NR_111510) was used as an out-group to root the tree [23]. Therefore due to the high degree of similarity we have named our strain as *Penicillium mallochii* LMB-HP37 (GenBank accession number KP979777). This species was first isolated from *Rothschildia lebeau* and *Citheronia lobesis* (*Saturniidae*) caterpillar guts and faeces, and leaves of *Spondias mombin* (*Anacardiaceae*) and it has been recently described and named by Rivera et al. [24].

In order to determine the cellulase production capacity of the strain, liquid cultures in shaken flasks were carried out with either lactose or microcrystalline cellulose as carbon sources at 28°C for 96 h and 168 h, respectively. Lactose was a better carbon source for growth and enzyme production (Fig. 3). Although there was a poor growth in the first 48 h, *P. mallochii* LMB-HP37 produced twice as much biomass on lactose than on cellulose (Fig. 3c, f) indicating that the fungus had to activate the genes related to lactose metabolism. It has been demonstrated that *P. chrysogenum* and other fungal species have a dual assimilation strategy for lactose, simultaneously employing extracellular and intracellular hydrolysis [25,26].

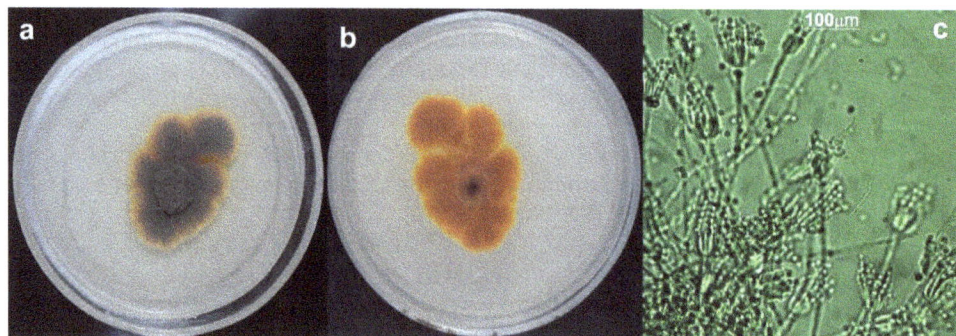

Fig. 1. Morphological characteristics of *Penicillium mallochii* LMB-HP37. PDA plates showing the obverse (a) and reverse (b) of fungal colony (c) microscopic observation

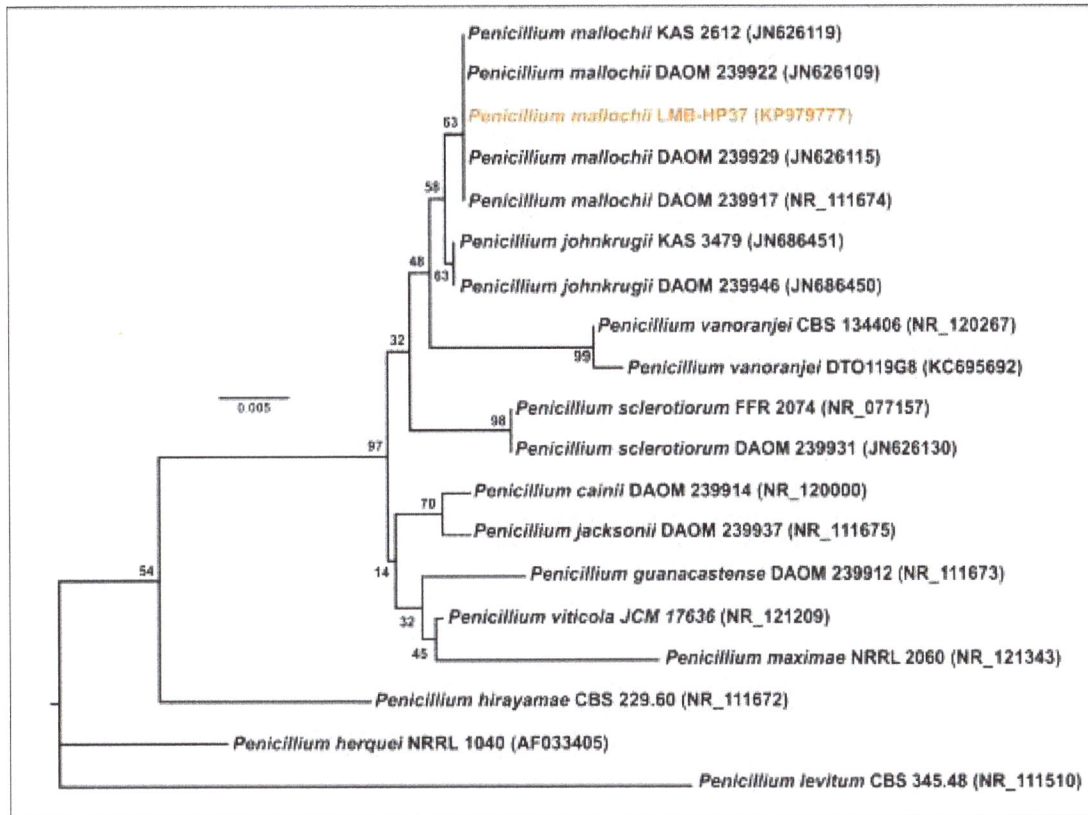

Fig. 2. Comparative sequence analysis of the ITS region of rRNA genes of *Penicillium mallochii* LMB-HP37

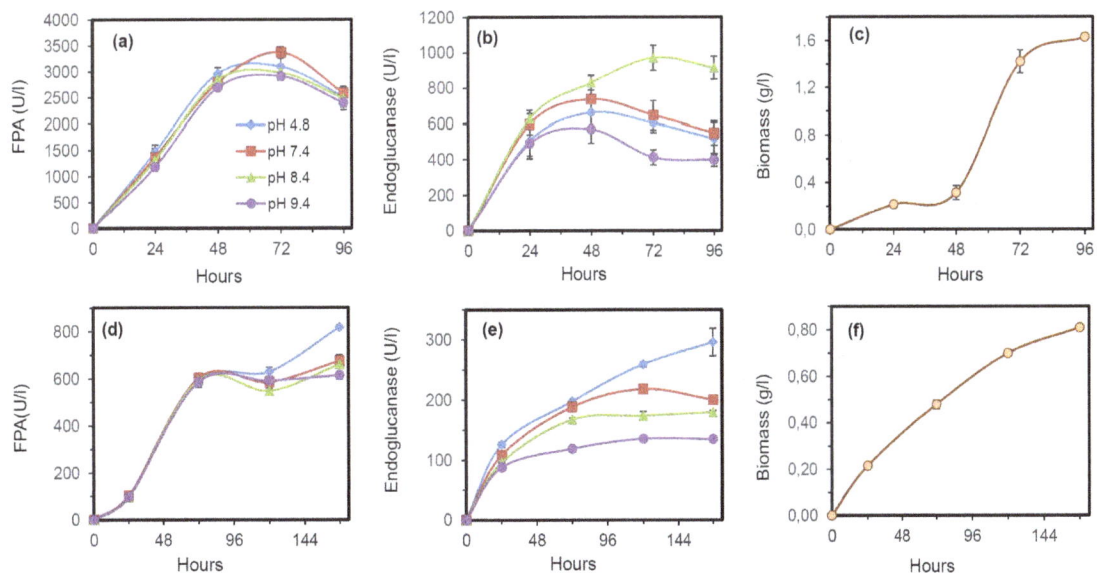

Fig. 3. Time course profiles of total cellulase (a, b), endoglucanase (b, e), and biomass (c, f) production by *Penicillium mallochii* LMB-HP37 on lactose (upper panel) and cellulose (bottom panel)

Commercial cellulase production is mainly due to the well known cellulase producer *Trichoderma reesei* and to *Aspergillus niger* to a lesser extent. Only few *Penicillium* species are described for cellulase production to a commercial level, i.e., *Penicillium funiculosum* [27]. Our strain *P. mallochii* LMB-HP37 produced high total cellulase titers (3096±179 U/l at the standard pH 4.8) on lactose in a shorter period of time (Fig. 3a) than most reported *Penicillium* species [27]. On cellulose however, it produced about four-fold less FPA (Fig. 3d) than on lactose, indicating that this sugar is a good cellulase inducer. Interestingly, much higher activity was attained for neutral cellulase on lactose (3371±108 U/l at pH 7.4) and alkaline cellulases attained similar activity levels than the acid cellulase (2978±151 U/l at pH 8.4 and 2910±42 U/l at pH 9.4). This indicates that *P. mallochii* LMB-HP37 is a true and good producer of alkaline cellulase. Although several *Penicillium* species have been found to be good cellulase producers very few have been reported for producing alkaline cellulases [28,29,30,31]. Comparing to other *Penicillium* species that are reported as good cellulase producers, our strain *P. mallochii* LMB-HP37 produced 1.3-fold more FPA than *P. funiculosum* and similar FPA levels than *P. janthinellum* [27], 2.1- to 2.5-fold more FPA than mutant strain *P. echinulatum* 9A02S1 [28,30] and similar levels of FPA when this mutant strain was grown in a bioreactor under fed-batch regime [29], and about 3.7-fold more FPA than *P. simplicissimum* TUB-F-2378 and *P. pulvillorum* TUB-F-2220 [31].

Endoglucanase production level by *P. mallochii* LMB-HP37 was less than total cellulase indicating that this strain probably produces more active cellobiohydrolases. However, as shown in Figs. 3b and 3e, endoglucanase activity at different pH values peaks also at different fermentation times indicating that they may be encoded by different genes. Five endoglucanases and two cellobiohydrolases have been found in *P. pinnophilum* [32]. Also, two cellobiohydrolases and three endoglucanases are produced by *P. brasilianum* IBT 20888 [33]. The highest endoglucanase activity produced by *P. mallochii* LMB-HP37 was found at pH 8.4 and the second major activity was at pH 7.4. *Penicillium citrinum* MTCC 6489 has been found to produce a neutral endoglucanase [17].

Technical parameters related to cellulase production by *P. mallochii* LMB-HP37 are depicted in Table 1. High specific FPA activity was attained on lactose at pH 7.4 (23.4±0.8 U/mg$_{protein}$) while maximum specific endoglucanase activity was obtained at pH 8.4 (4.5±0.6 U/mg$_{protein}$) [27,29,34]. Also, FPA and endoglucanase activities were produced at high volumetric (46.8±1.5 and 13.5±1.0 U/l.h, respectively) and specific (32.9±1.1 and 9.5±0.7 U/g$_{biomass}$.h, respectively) productivities at the same pHs which indicate that this strain may be suitable for commercial development. As referred before, lactose was a better carbon source for cellulase production than cellulose. Lactose may be a better inducer for the expression of cellulase genes as it has been found in *P. decumbens* [35]. In addition, lactose is a cheap carbon source produced as byproduct in dairy plants which is important for the economy of a cellulase production process.

Table 1. Comparison of some technical parameters for total cellulase and endoglucanase production on lactose liquid cultures of *Penicillium mallochii* LMB-HP37 at different pH

Parameter	pH 4.8	pH 7.4	pH 8.4	pH 9.4
FPA	3096±179[ab]	3371±108[a]	2978±151[b]	2910±42[b]
Y$_{FPA/x}$	2178±126[ab]	2371±76[a]	2095±107[b]	2047±30[b]
Y$_{FPA/prot}$	21525±1245[ab]	23438±754[a]	20709±1053[b]	20236±294[b]
Γ$_{FPA}$	42.9±2.5[ab]	46.8±1.5[a]	41.4±2.1[b]	40.4±0.6[b]
q$_{FPA}$	30.2±1.8[ab]	32.9±1.1[a]	29.1±1.5[b]	28.4±0.4[b]
EG	604±53[b]	647±82[b]	971±70[a]	413±42[c]
Y$_{EG/x}$	425±38[b]	455±57[b]	683±49[a]	291±30[c]
Y$_{EG/prot}$	4200±370[b]	4500±566[b]	6750±487[a]	2871±293[c]
Γ$_{EG}$	8.4±0.7[b]	9.0±1.1[b]	13.5±1.0[a]	5.7±0.6[c]
q$_{EG}$	5.9±0.5[b]	6.3±0.8[b]	9.5±0.7[a]	4.0±0.4[c]

FPA = Total cellulase activity (U/l); Y$_{FPA/x}$ = Total cellulase specific activity (U/g$_{biomass}$); Y$_{FPA/Prot}$ = Total cellulase specific activity (U/g$_{protein}$); Γ$_{FPA}$ = Total cellulase volumetric productivity (U/l.h); q$_{FPA}$ = Total cellulase specific productivity (U/g$_{biomass}$.h); EG = endoglucanase activity (U/l); Y$_{EG/x}$ = Endoglucanase specific activity (U/g$_{biomass}$); Y$_{EG/Prot}$ = Endoglucanase specific activity (U/g$_{protein}$); Γ$_{EG}$ = Endoglucanase volumetric productivity (U/l.h); q$_{EG}$ = Endoglucanase specific productivity (U/g$_{biomass}$.h). Values represent mean of four repetitions±SD. Means with the same letter within a row are no significantly different (P < .01). Data were evaluated by ANOVA and Duncan's test

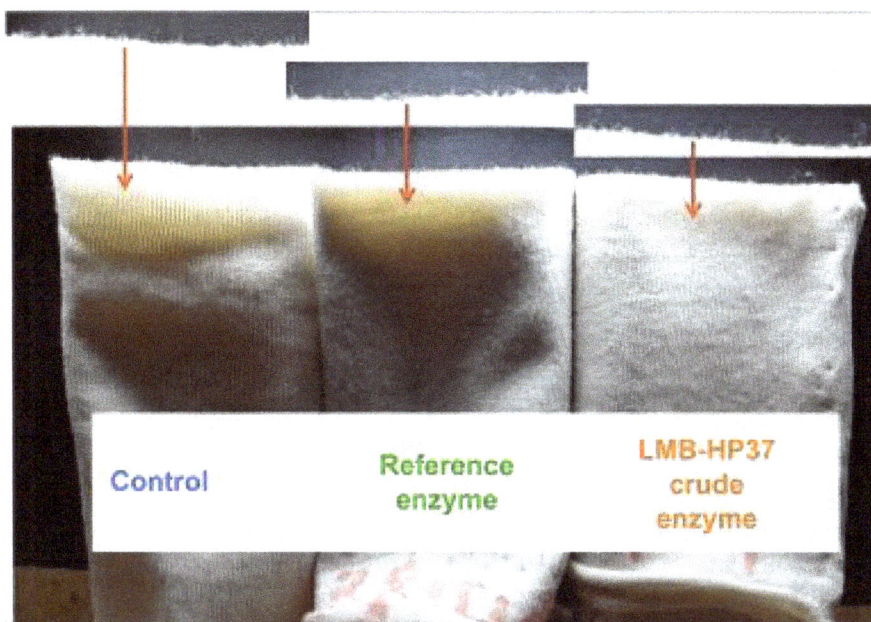

Fig. 4. Anti-pilling assessment of *Penicillium mallochii* LMB-HP37 cellulase

Textile industry is probably the main user of cellulases for several processes like the removal of surface fiber fibrils from cellulosic fabrics to avoid pilling and improve fabric appearance and the ageing of indigo-dyed denim garments [3,5,11]. In this sense, we have tested the activity of *P. mallochii* LMB-HP37 culture filtrates to reduce pilling as compared with a commercial enzyme preparation at pH 7 and similar cellulase activities. As shown in above Fig. 4 above, *P. mallochii* LMB-HP37 enzyme had slightly better results as an anti-pilling agent even though is a crude preparation. Moreover, *P. mallochii* LMB-HP37 enzyme still works at higher pH values at which commercial enzyme did not work. This enzyme preparation may be also used for demin biostoning and bioblasting. Considering that *P. mallochii* LMB-HP37 is a wild strain there is room for its genetic improvement as well as for process optimization [36].

4. CONCLUSION

Alkaline cellulases are demanded by the textile industry for several purposes but commercial preparations showing activity at alkaline conditions are very scarce. A previously isolated strain that showed alkaline cellulase activity has been molecularly identified as *Penicillium mallochii* LMB-HP37 and further characterized for cellulase production, being this the first report on its capability for producing alkaline cellulases

active up to pH 9.4. This strain produced high total cellulase activity on lactose which compares to well-known cellulase producers but at neutral to alkaline pH levels. Moreover, this strain produces high endoglucanase activity at pH 8.4 and 9.4. Data obtained reveal that the crude enzyme is suitable for anti-pilling process (biopolishing) and may be also useful for biostoning.

ACKNOWLEDGEMENTS

This research was supported by grant N0 109-FINCyT-FIDECOM-PIPEA-2012 from the National Program of Innovation for Competitiveness and Productivity of Peru. The authors wish to thank to Mrs. Mary Pasmiño for her technical assistance and to Mr. Gianangelo Nava of Química Nava, S.A.C for helping with the anti-pilling assessment.

COMPETING INTERESTS

Authors have declared that no competing interests exist.

REFERENCES

1.   Nigam PS. Microbial enzymes with special characteristics for biotechnological applications. Biomolecules. 2013;3(3):597-611.

2. Adrio JL, Demain AL. Microbial enzymes: Tools for biotechnological processes. Biomolecules. 2014;4(1):117-139.

3. Araújo R, Casal M, Cavaco-Paulo A. Application of enzymes for textile fibres processing. Biocatal Biotransform. 2008; 26(5):332-349.

4. Soares JC, Moreira PR, Queiroga AC, Morgado J, Malcata FX, Pintado ME. Application of immobilized enzyme technologies for the textile industry: A review. Biocatal Biotransform. 2011;29(6): 223–237.

5. Durán N, Durán M. Enzyme applications in the textile industry. Rev Prog Coloration. 2000;30(1):41-44.

6. Tzanov T, Calafell M, Guebitz GM, Cavaco-Paulo A. Bio-preparation of cotton fabrics. Enzyme Microb Technol. 2001; 29(6-7):357–362.

7. Zhang Y-HP, Himmel ME, Mielenz JR. Outlook for cellulase improvement: Screening and selection strategies. Biotechnol Adv. 2006;24(5):452-81.

8. Wilson DB. Cellulases and biofuels. Curr Opin Biotechnol. 2009;20(3):295-9.

9. Kuhad RC, Gupta R, Singh A. Microbial cellulases and their industrial applications. Enzyme Res. 2011;2011: Article ID 280696, DOI:10.4061/2011/280696.

10. Juturu V, Wu JC. Microbial cellulases: Engineering, production and applications. Renew Sust Energ Rev. 2014;33:188-203.

11. Shah SR. Chemistry and applications of cellulase in textile wet processing. Res J Engineering Sci. 2014;3(2):1-5.

12. Anish R, Rahman, MS, Rao M. Application of cellulases from an alkalothermophilic Thermomonospora sp. in biopolishing of denims. Biotechnol Bioeng. 2007;96(1): 48–56.

13 Miettinen-Oinonen A, Londesborough J, Joutsjoki V, Lantto R, Vehmaanperä J. Three cellulases from Melanocarpus albomyces for textile treatment at neutral pH. Enzyme Microb Technol. 2004; 34(3-4):332-341.

14. Fujinami, S, Fujisawa, M. Industrial applications of alkaliphiles and their enzymes – past, present and future. Environ Technol. 2010;31(8-9):845-856.

15. Landaud. S., Piquerel P, Pourquié J. Screening for bacilli producing cellulolytic enzymes active in the neutral pH range. Lett Appl Microbiol. 1995;21(5):319-321.

16. Ito S. Alkaline cellulases from alkaliphilic Bacillus: Enzymatic properties, genetics, and application to detergents. Extremophiles. 1997;1(2):61-66.

17. Dutta T, Sahoo R, Sengupta R, Ray SS, Bhattacharjee A, and Ghosh S. Novel cellulases from an extremophilic filamentous fungi Penicillium citrinum: Production and characterization. J Ind Microbiol Biotechnol. 2008;35(4):275–282.

18. Vega K, Villena GK, Sarmiento VH, Ludeña Y, Vera N, Gutiérrez-Correa M. Production of alkaline cellulase by fungi isolated from an undisturbed rain forest of Peru. Biotechnol Res Int. 2012;2012. Article ID 934325.
DOI: 10.1155/2012/934325.

19. Möller EM, Bahnweg G, Sandermann H, Geiger HH. A simple and efficient protocol for isolation of high molecular weight DNA from filamentous fungi, fruit bodies and infected plant tissues. Nucl Acids Res. 1992;20(22):6115-6116.

20. Bellemain E, Carlsen T, Brochmann C, Coissac E, Taberlet P, Kauserud H. ITS as an environmental DNA barcode for fungi: An in silico approach reveals potential PCR biases. BMC Microbiology. 2010; 10:189. DOI: 10.1186/1471-2180-10-189.

21. Ghose TK. Measurement of cellulolytic activities. Pure Appl Chem. 1987;59(2): 257–258.

22. Lowry OH, Rosebrough NJ, Farr AL, Randall RJ. Protein measurement with the folin phenol reagent. J Biol Chem. 1951; 193(1):265-275.

23. Visagie CM, Houbraken J, Rodriques C, Pereira CS, Dijksterhuis J, Seifert KA, Jacobs K, Samson RA. Five new Penicillium species in section Sclerotiora: A tribute to the Dutch Royal family. Persoonia. 2013;31:42–62.

24. Rivera KG, Díaz J, Chavarría-Díaz F, Garcia M, Urb M, Thorn RG, Louis-Seize G, Janzen DH, Seifert KA. Penicillium mallochii and P. guanacastense, two new species isolated from Costa Rican caterpillars. Mycotaxon. 2012;119:315-328.

25. Jonás A, Fekete E, Flipphi M, Sándor E, Jäger S, Molnár AP, Szentirmai A, Karaffa L. Extra- and intracellular lactose catabolism in Penicillium chrysogenum: Phylogenetic and expression analysis of the putative permease and hydrolase genes. J Antibiot. 2014;67(7):489-497.

26. Seiboth B, Pakdaman BS, Hart L, Kubicek CP. Lactose metabolism in filamentous fungi: How to deal with an unknown

substrate. Fungal Biol Rev. 2007;21(1): 42-48.

27. Gusakov AV, Sinitsyn AP. Cellulases from *Penicillium* species for producing fuels from biomass. Biofuels. 2012;3(4):463–477.

28. Sehnem NT, de Bittencourt LR, Camassola M, Dillon AJ P. Cellulase production by *Penicillium echinulatum* on lactose. Appl Microbiol Biotechnol. 2006; 72:163-167.

29. Pereira BMP, Alvarez TM, Delabona PdaS, Dillon AJP, Squina FM, Pradella JGdaC. Cellulase on-site production from sugar cane bagasse using *Penicillium echinulatum*. Bioenerg Res. 2013;6(3): 1052-1062.

30. Camassola M, Dillon AJP. Effect of different pretreatment of sugar cane bagasse on cellulase and xylanases production by the mutant *Penicillium echinulatum* 9A02S1 grown in submerged culture. Bio Med Res Int. 2014;2014: Article ID 720740.
DOI: 10.1155/2014/720740.

31. Marjamaa K, Toth K, Bromann PA, Szakacs G, Kruus K. Novel *Penicillium* cellulases for total hydrolysis of lignocellulosics. Enzyme Microb Technol. 2013;52(6-7):358–369.

32. Wood TM, McCrae SI, Bhat KM. The mechanism of fungal cellulase action: Synergism between enzyme components of *Penicillium pinophilum* cellulase in solubilizing hydrogen bond-ordered cellulose. Biochem J. 1989;260(1):37-43.

33. Jørgensen H, Eriksson T, Börjesson J, Tjerneld F, Olsson L. Purification and characterization of five cellulases and one xylanase from *Penicillium brasilianum* IBT 20888. Enzyme Microb Technol. 2003; 32(7):851–861.

34. Jeya M, Joo AR, Lee KM, Sim WI, Oh DK, Kim YS, Kim IW, Lee JK. Characterization of endo-β-1,4-glucanase from a novel strain of *Penicillium pinophilum* KMJ601. Appl Microbiol Biotechnol. 2010;85(4): 1005–1014.

35. Wei X, Zheng K, Chen M, Liu G, Li J, Lei Y, Qin Y, Qu Y. Transcription analysis of lignocellulolytic enzymes of *Penicillium decumbens* 114-2 and its catabolite-repression-resistant mutant. CR Biologies. 2011;334(11):806–811.

36. Singhania RR, Sukumaran RK, Patel AK, Larroche C, Pandey A. Advancement and comparative profiles in the production technologies using solid-state and submerged fermentation for microbial cellulases. Enzyme Microb Technol. 2010; 46(7):541–549.

5

Establishment of Callus Derived from *Jatropha curcas* L. Petiole Explants and Phytochemical Screening

K. Nyembo[1*], N. Mbaya[1], N. Muambi[1] and C. Ilunga[2]

[1]*Department of Phytobiology, Division of Life Science, General Atomic Energy Commission, Regional Center of Nuclear Studies of Kinshasa XI, University of Kinshasa, DR, Congo.*
[2]*Department of Biotechnology, Division of Life Science, General Atomic Energy Commission, Regional Center of Nuclear Studies of Kinshasa, XI, University of Kinshasa, DR, Congo.*

Authors' contributions

This work was carried out in collaboration between all authors. Author KN designed the study, wrote the protocol and the first draft of the manuscript. Author NM performed critical reviews of the manuscript. Authors NM and CI conducted the experiments of the study. All authors read and approved the final manuscript.

Editor(s):
(1) Standardi Alvaro, Department of Agricultural and Environmental Sciences, University of Perugia, Italy.
(2) Kuo-Kau Lee, Department of Aquaculture, National Taiwan Ocean University, Taiwan.
Reviewers:
(1) Anonymous, Brazil.
(2) Ihsan Ullah, Kyungpook National University, South Korea.
(3) Anonymous, Malaysia.
(4) Anonymous, Saudi Arabia.
(5) Anonymous, Nigeria.

ABSTRACT

Aims: The focus of this study was to assess the production of secondary metabolites of callus derived from *Jatropha curcas* L. petiole explants.
Study Design: Laboratory experimental tests; Tissue culture, Lyophilisation, Phytochemical Analysis, Determination of tannins by Folin-Denis method.
Place and Duration of the Study: Department of Phytobiology, Department of Biotechnology. General Atomic Energy Commission. Regional Center of Nuclear Studies of Kinshasa P.O BOX 868 Kin.XI DR Congo during March and June 2013.
Methodology: *In vitro* callus cultures were initiated from petiole explants of *Jatropha curcas* L. on

Corresponding author: Email: baudyne@yahoo.fr

Murashige & Skoog (1962) basal medium supplemented with growth regulator formulations 4.44 µM of 6-benzylaminopurine (BAP), 4.92 µM of Indole-3-butyric acid (IBA), and 100 ml L^{-1} coconut milk. Callus was lyophilized. The extracts were subjected to phytochemical tests for the presence of plant secondary metabolites. The amount of tannins was estimated by Folin-Denis method.

Results: Excellent growth of callus was obtained. Callus was soft, friable, lush green in color and grew fast from 8 to 30 days of culture then stabilized at low growth rate. The results obtained by phyto-chemical tests revealed the presence of saponins, tannins, alkaloids, steroids and flavonoids. The tannin contents *in vitro* proliferated callus, leaves, stem barks, root barks, roots and latex of *J. curcas* were found to be 161, 173, 220, 214, 43 and 245 µg tannic acid equivalent/g of dry weight respectively.

Conclusion: The results of this study have revealed that *in vitro* induced callus from *J. curcas* petiole explants was able to produce secondary metabolites. The pharmacological activity of *J. curcas* reported by various researchers can be attributed to the presence of these phytochemicals.

Keywords: Callus; in vitro culture; phytochemical; tannin; secondary metabolites.

1. INTRODUCTION

Jatropha curcas L. (*J. curcas*) belongs to *Euphorbiaceae* family and grows in tropical and sub-tropical regions. This crop has gained lot of importance due to property to produce biodiesel [1]. Furthermore, *J. curcas* has been used as traditional medicine to cure various infections and extracts from various parts of the plant, such as seeds, stem barks, roots and leaves have been shown fungicidal [2,3] and bactericidal properties [4-7].

In vitro culture techniques for *J. curcas* have been researched and developed in many laboratories for potential mass propagation [8-12]. Cell suspension culture is considered one of the best approaches for studying the biosynthesis of natural products, and callus is the richest sources of cell mass when establishing such cultures [13]. The past few decades have seen increasing scientific interest in both growth of modern plant tissue culture and the commercial development of this technology as means of producing valuable phyto chemicals [14].

Phenolic compounds which include tannins, phenolic acids and flavonoids are considered to be an important class of phytochemicals and were found in high concentrations in a wide variety of plant. These compounds possess many biological effects which are mainly attributed to their strong antioxidant properties in scavenging free radicals inhibition of peroxidation and chelating transition metals [15]. Free radicals can be harmful to the organism if not properly regulated and thus may cause several degenerative diseases, such as diabetes, aging, cancer, cardiovascular diseases, metabolic

syndrome and arthrosclerosis [16]. Antioxidants can protect the human body from frees radicals, reactive oxygen species effects and reduce the risk of cardiovascular and neurodegenerative disease [17-19].

The specific objective of this tissue culture study was to induce callus from petiole explants of *J. curcas* and to assess the production of secondary metabolites.

2. MATERIALS and METHODS

2.1 Collection of Plant Material

Fresh leaves, petioles, roots, root barks, stem barks, latex of *J. curcas* were collected from experimental garden of CREN-K /Kinshasa RD-Congo.

2.2 Initiation of Callus and Maintenance

Starting materials of 150 petioles were used as initial explants for callus production. Under the laminar airflow cabinet, we cut them of 3-4 mm and then they were sterilized with 0.25% mercuric chloride solution for 4 minutes followed by 6 rinses in sterile distilled water to remove the traces of the sterilizing substance. Five petiole explants were transferred aseptically on the surface of a solidified medium in each Petri dish containing 25 ml of macro and micro salts Murashige & Skoog, (MS) [20] supplemented with 4.44 µM Benzylaminopurine (BAP), 4.92 µM Indole-3-butyric acid (IBA) and 100 ml/L coconut milk for initiation of callus. The pH of the medium was adjusted to 5.8 using NaOH 0.1N and Difco agar (0.8% w/v) was used as solidifying agent before autoclaving at 120°C for 20 minutes.

Cultured Petri dishes were incubated for callus formation in a growth room at 27±0, 1°C under a 16 hours photoperiod with a light intensity of 2000 lux. Calli were sub-cultured after 14 days interval on fresh medium for further proliferation period of two months.

2.3 Lyophilization

The process of freeze-drying was done using the freeze-dryer 7670530 LABCONCO. This method can protect biological activity of samples. Callus and latex were frozen in vials at a low temperature (-30°C). The water was then extracted via vacuum during 24 hours, resulting in a porous dry "lyocake". Lyophilized callus and latex were pulverized and sieved to 0.2 mm mesh. These materials can be easily stored.

2.4 Preparation of Extracts for Phytochemical Analysis

The dried samples were pulverized to fine powder using an electric grinder. Five grams of each powdered plant material were submitted to cold extraction in analytical grade methanol and kept at room temperature for 24 hrs. Extracts were recovered by filtration using Whatman N° 1 filter paper. Excess of methanol was evaporated in hot-air oven at 45°C and later the condensed extract was weighed and reconstituted in minimum volume of methanol and stored at 4°C for further analysis.

2.5 Phytochemical Analysis of the Plant Extract

The methanolic extracts were subjected to preliminary qualitative phytochemical screening for the presence of plant secondary metabolites such as alkaloids, saponins, steroids, tannins and flavonoids using methods described by Harborne [21] and Evans [22].

2.6 Determination of tannins by Folin-Denis Method

The amount of tannins was estimated by Folin-Denis method. Tannic acid was taken as the standard for estimating the total amount of tannins in the methanolic extract. The tannin contents were calculated as tannic acid equivalent from the calibration curve of standard tannic acid by plotting the absorbance versus concentration. Different dilutions of standard tannic acid were prepared ranging (0 g/l, 0.5 g/l, 1 g/l, 1.5 g/l, and 2 g/l). This was then transferred to two sets of tubes. These tubes were then

incubated at 80°C in a water bath during one hour. The absorbance was measured with a spectrophotometer at 760 nm wavelength. The tannin content was expressed in terms of tannic acid equivalents/g dry weight (TAE/g DW).

3. RESULTS AND DISCUSSION

3.1 Callus Formation

Eighty five per hundred of the total petiole explants cultured attained the stage of callus formation approximately 8 days after incubation. According to our visual observations, callus arising from the edges of explants was soft and friable. The callogenesis rate was fast from 8 until 30 days of culture, and then stabilized at a low growth rate (Figs. 1 and 2). Green points were observed on the surface of this callus which was maintained by transferring on the same fresh medium after every two weeks during two months.

Fig. 1. Initiation of callus formation from petiole explants

Fig. 2. Proliferated callus from petiole explants

Hypocotyl explants of *J. curcas* cultured in MS medium supplemented with 2.26 μM 2.4-D alone and with 2% v/v coconut milk proved to be more effective for establishment of callus on a large scale in short period of time [23]. Although, it has been reported that 2.4-D in low concentration is the most commonly used auxin for good callogenesis, in the present study best callus growth was found when petiole explants were placed on full strength MS medium supplemented with a 1:1 concentration ratio of BAP and IBA with 100ml/L coconut milk. Gawri and Upadhy [24] reported that tannins, alkaloids, glycosides, flavonoids, and phlobatannis were present in the petioles of fresh leaves of *Jatropha curcas* as well as the cultured callus but the concentration of alkaloids and glycosides were higher in the callus. It is well known that, *in vitro* cultures are able to produce secondary metabolites in quantities more than that of original parts of the plants [25].

3.2 Phytochemical Analysis

The qualitative estimation of five secondary metabolites (saponins, tannins, alkaloids, steroids and flavonoids) was undertaken. The phytochemical investigations showed that all secondary metabolites analyzed were present in the plant parts as well as the callus (Table 1). Using qualitative analysis, Igbinosa et al. [4], Sudha et al. [7] and Asobara et al. [26] have reported the presence of the same compounds in all tissues of *J. curcas* studied: stem bark, leaves, roots and seeds extracts. All the phytochemical components detected were known to support bioactive activities in medicinal plants [27,28]. These bioactive molecules are produced in plants in less quantity, which is largely influenced by environmental factors. This has led to looking for alternate resources like callus and suspension cultures where the synthesis of bioactive molecules can be enhanced either with PGRs or elicitors [29].

3.3 Determination of Phenolic Compounds

The presence of tannins was determined by Folin-Denis method. This is based on the reduction of phosphomolybdic acid and tungstic in alkaline medium in the presence of tannins to give a blue color whose intensity is measured at 760 nm. In Table 2 is reported that the tannin contents of callus, leaves, stem barks, root barks, roots and latex were found to be 161, 173, 220, 214, 43 and 245 µg TAE/g DW respectively. In the current study, highest and lowest tannin contents of 245 and 43µg TAE/g DW were recorded in latex and roots methanolic extract

respectively. Previous studies have reported the presence of tannins in *J. curcas* latex [30,31]. Igbinosa et al. [32] have found a high polyphenolic content in *J. curcas* stem bark. The tannins possess anticancer activity and they can be used in cancer prevention [33]. Tannins found in the plant are useful in the treatment of inflamed or ulcerated tissues. They have also been used for treating intestinal disorders such as diarrhea and dysentery [34]. The presence of tannins in *J. curcas* supports the traditional medicinal use of this plant in the treatment of different human ailments.

Table 1. Qualitative analysis of methanolic extracts of *J. curcas*

Compounds	*J. curcas*		
	Callus	Leaves	Root barks
Saponins	+	+	+
Tannins	+	+	+
Alcaloïds	+	+	+
Stéroïds	+	+	+
Flavonoïds	+	+	+

Note: '+' indicates presence

Table 2. The tannin contents of the methanolic extracts of *J. curcas*

J. curcas	Concentrationµg TAE/g DW
Callus	161
Leaves	173
Stem barks	220
Root barks	214
Roots	43
Latex	245

4. CONCLUSION

The results of this study have revealed that *in vitro* induced callus from *Jatropha curcas* L. petiole explants was able to produce secondary metabolites. Petioles were suitable explant material. The concentration of tannin in callus was observed to be 161 µg TAE/g DW. *In vitro* culture should be used to increase the production of secondary metabolites. The presence of bioactive compounds in the plant parts of *J. curcas* such as saponins, tannins, alkaloids, steroids and flavonoids indicates the medicinal values of this plant. Therefore, further research is needed for the isolation, identification and quantification of other biopharmaceuticals in the *J. curcas* extracts.

ACKNOWLEDGEMENTS

The authors express their sincere thanks to Mr. DG. Kombele for his help in reviewing the manuscript.

COMPETING INTERESTS

Authors have declared that no competing interests exist.

REFERENCES

1. Abdulla R, Chan ES, Ravindra P. Biodiesel production from *Jatropha curcas*: A critical review. Crit. Rev. in Biotech. 2011;31:53-64.

2. Saetae D, Worapot S. Antifungal activities of ethanolic extract from *Jatropha curcas* seed cake. J Microbiol Biotechnol. 2010;20(2):319-24.

3. Chandel U, Pimpalgaonkar R. Efficacy of leaf exudate of *Jatropha curcas* L. on percentage spore germination inhibition of its selected phylloplane and rhizosphere fungi. Indian J Sci Res. 2014;4(1):70-4.

4. Igbinosa OO, Igbinosa EO, Aiyegoro OA. Antimicrobial activity and phytochemical screening of stem bark extracts from *Jatropha curcas* (Linn). African J Pharmacol. 2009;3(2):58-62.

5. Gupta SM, Arif M, Ahmed Z. Antimicrobial activity in leaf, seed extract and seed oil of *Jatropha curcas* L. J Appl Nat Sci. 2011;3(1):102-5.

6. Nyembo K, Kikakedimau N, Mutambel' H, Mbaya N, Ekalakala T, Bulubulu O. *In vitro* antibacterial activity and phytochemical screening of crude extracts from *Jatropha curcas* Linn. Eur J Med Plants. 2012;2(3):242-51.

7. Sudha B,Dolly M, Rupali R, Neha A, Sangeeta S. Ecofriendly finishing of fabric with *Jatropha curcas* leaves. Res J Family Community and Consumer Sci. 2013;1(1):7-9.

8. MedzaMvé SD, Mergeai G, Baudoin JP, Toussaint A. Amélioration du taux de multiplication *in vitro* de *Jatropha curcas* L. Tropicultura. 2010;28(4) :200-4.

9. Kumar N, Reddy MP. Thidiazuron (TDZ) induced plant regeneration from cotyledonary petiole explants of elite genotypes of *Jatropha curcas*: A candidate biodiesel plant. Ind. Crop. Prod. 2012;39:62–8.

10. Nahar K,Borna RS. *In vitro* plant regeneration from shoot tip explants of *Jatropha curcas* L: A. J Sci Technol. 2013;3(1):38-42.

11. Renuga LG, Rajamanickam C. Efficient *In vitro* regeneration of biodiesel plant *Jatropha curcas*. J Plant Agri Res. Art. ID: PAR14 01. 2014;1(1):1-6.

12. Franco MC, Marques D, Siqueira WJ, Latado RR. Micropropagation of *Jatropha curcas* superior genotypes and evaluation of clonal fidelity by target region amplification polymorphism (TRAP) molecular marker and flow cytometry. African J Biotechnol. 2014;13(38):3872-80.

13. Edeoga HO, Okwu DE, Mbaebie BO. Phytochemical constituents of some Nigerian medicinal plants. African J Biotechnol. 2005;4:685-8.

14. Havkin-Frankel D, Dorn R, Leustek T. Plant tissue culture for production of secondary metabolites. Food Technol. 1997;51:56-61.

15. Bahman N, Mohammed K, Hamidreza I. *In vitro* free radical scavenging activity of five salvia species. Pak J Pharm Sci. 2007;20(4):291-4.

16. Dröge W. Free radicals in the physiological control of cell function. Physiol Rev. 2002;82:47–95.

17. Hang B Huifeng R, Hidak IE, Yukihiko T, Testuhito H. Effects of heating and the addition of seasonings on the anti-mutagenic and anti-oxidative activities of polyphenols. Food Chem. 2004;86(4):517-24.

18. Manach C, ScalbertA, Moeand C, Remsy C, Jimenez L. Polyphenols: Food sources and bioavailability. Am J Clin Nutr. 2004;79:727-47.

19. Gülcin I. Antioxidant activity of caffeic acid. Toxicology.2006; 217:213–20.

20. Murashige T, Skoog F. A revised medium for rapid growth and bioassays with tobacco tissue cultures. Physiol. Plant. 1962;15:473–79.

21. Harborne JB. Phytochemical Methods: A guide to modern techniques of plants analysis. 3rd ed. London: Chapman and Hall. 1998;182-190.

22. Evans WC. Evans and Trease Pharmacognosy. 14th ed. WB Saunders Company Ltd. 1996;224-93.

23. Soomro R, Memon RA. Establishment of callus and suspension culture in *Jatropha curcas*. Pak J Bot. 2007;39:2431-41.

24. Gawri S, Upadhyay A. A comparative study on the antimicrobial activity and the presence of phytochemicals in the petioles and callus of *J. curcas*. J Phytol. 2012;4(3):18-20.

25. Namba T, Morita O, Huang SL, Goshima K, Hottori M, Kakuchi N. Studies on cardio-active crude drugs. Effect of coumarins on cultured myocardial cells. Planta Medica. 1988;54:277-82.

26. Harry-Asobara JL, Linda J, Eno-Obongkon S. Comparative study of the phytochemical properties of *Jatropha curcas* and *Azadirachta indica* plant extracts. J Poison Med Res plants. 2014;2(2):20-4.

27. Fugitha Y, Hara Y, Ogino T, Suga C. Production of shikonin derivatives by cell suspension cultures of *Lithospermum erythrihizon*. Plant Cell Rep. 1981;1:59-60.

28. Sirigiri CK, Kokkanti M, Patchala A. An efficient protocol devised for rapid callus induction from leaf explants of *Biophytum sensitivum* (Linn) DC. Int J Phytopharm. 2014;4(1):20-4.

29. White PR. Potentially unlimited growth of excised plant callus in an artificial nutrient. Am J Bot. 1939;26:59-64.

30. Perry LM, Metzger J. Medicinal plants of East and Southeast Asia: Attributed properties and uses. Cambridge: 29 the MIT Press. 1980;246-7.

31. Ievens M, Vanden-Berghe DA, Mertens F, Vlietinck A, Lammens E. Screening of higher plants for biological activity. I. Antimicrobial activity. Planta Med. 1979;36: 311-2.

32. Igbinosa OO, Igbinosa IH, Chigor VN, Uzunuigbe OE, Oyedemi SO, Odjadjare EE, Okoh AI, Igbinosa EO. Polyphenolic contents and antioxidant potential of stem bark extracts from *Jatropha curcas* L. Int J Mol Sci. 2011;12:2958-71.

33. Li H, Wang Z, Liu Y. Review in the studies on tannins activity of cancer prevention and anticancer. Zhong Yao Cai. 2003; 26(6):444-8.

34. Dharmananda S. Gallnuts and the Uses of Tannins in Chinese Medicine. In: Proceedings of institute for Tradit. Med, Portland, Oregon; 2003.

35. Available:http://www.itmonline.org/arts/gall nuts.htm

6

Starch Phosphorylase: An Overview of Biochemical Characterization, Immobilization and Peptide Mapping

Ritu Jain[1], Sarika Garg[2] and Anil Kumar[1*]

[1]School of Biotechnology, Devi Ahilya University, Khandwa Road, Indore-452001, India.
[2]Department of Neuroscience, The University of Montreal Hospital Research Centre (CRCHUM), Bureau/Office: CRCHUM R09.476, Tour Viger, 900, rue St-Denis, Montréal, Québec H2X 0A9, Canada.

Authors' contributions

This work was carried out in collaboration between all authors. Authors RJ and SG have equal contribution in this manuscript. Author RJ prepared the first draft of the manuscript. Authors RJ and SG both searched literature. The manuscript has been finalized by authors SG and AK. Author AK also designed the study and managed the analyses of the study. All authors read and approved the final manuscript.

Editor(s):
(1) Samuel Peña-Llopis, Developmental Biology and Internal Medicine, Oncology Division, University of Texas Southwestern Medical Center, USA.
Reviewers:
(1) Anonymous, University of Genoa, Italy.
(2) Anonymous, Kagoshima University, Japan.
(3) Muhammad Yasin Naz, Dept. of Fundamental & Applied Sciences, Universiti Teknologi Petronas, 31750 Tronoh, Malaysia.
(4) Anonymous, Center for Research for Food and Development, México.
(5) Ana Lucia Ferrarezi, Department of Biology, Unesp, Brazil.

ABSTRACT

Starch phosphorylase (EC 2.4.1.1) is the key enzyme which catalyzes the reversible conversion of starch and inorganic phosphate into glucose-1-phosphate (G-1-P) and has been reported in many higher plants. It can be exploited for industrial purposes for tailoring starch and for the synthesis of biopolymers which may be useful in food industry and are also of clinical importance. Starch in the form of insoluble granules accumulates in chloroplasts as the primary product of photosynthesis. Multiple forms of the enzyme have been reported in different plant tissues which have been predicted to have different roles in starch metabolism. Here, various biochemical properties have

Corresponding author: Email: ak_sbt@yahoo.com

been reviewed. Various techniques for enzyme immobilization have been discussed. Besides, reports on immobilization of starch phosphorylase from different sources and on different solid matrices have also been reviewed. Peptide mapping reports of various proteins have also been assessed in this review.

Keywords: Starch phosphorylase; glucose-1-phosphate; biochemical aspects; immobilization; peptide mapping.

1. INTRODUCTION

The glucans are the polysaccharides which have monomers of D-glucose linked by glycosidic bond in their structure. There are two types of glucans- α and β. The α – glucans include dextran, glycogen, starch and pullulan and β-glucans include cellulose, curdlan, laminarin, chrysolaminarin, lentinan, pleiran and zymosan. The family Glycosyl transferases, is the largest family of the enzymes of carbohydrate metabolism. Polyglucans (starch and glycogen) are the main source of carbon and energy for many organisms and α-glucan phosphorylase plays an important role in glucan metabolism and has wide distribution in plants, animals, and prokaryotes. Its mode of action is degrading polyglucans by phosphorolysis and is mostly active in the homodimeric form with vitamin cofactor, pyridoxal 5'-phosphate [1,2]. Starch acts as primary energy source or reserve in higher plants. Starch in the form of insoluble granules accumulates in chloroplasts as the primary product of photosynthesis.

Amylose and Amylopectin are the basic components of starch (Fig. 1). ADP-glucose pyrophosphorylase, starch synthase and branching enzyme play a major role in starch synthesis. Amyloplast is the organelle in plant storage cells/tissues where starch is synthesized and deposited. Biosynthesis of starch or starch like polymers in plants with functional properties can be obtained by using different techniques like knockout, antisense RNA techniques and by expressing heterologous genes [3]. In developing seeds, starch production takes place and this starch acts as a carbon reservoir or storage for long term usage [4]. At the simplest level, the process of starch degradation requires an initial hydrolytic attack on the intact starch granule followed by de-branching (hydrolysis) of α - (1, 6)-linkages to produce linear glucan chains, and finally, the degradation of the linear chains to glucosyl monomers. There is a range of plastidial enzymes with starch-degrading capabilities which may participate in the process of starch degradation and turnover. The α -amylase (EC

3.2.1.1) and other endo-amylases hydrolytically cleave α -(1-4)-glucosyl bonds resulting in the production of a mixture of linear and branched moieties and, ultimately, glucose, maltose, malto-triose, and a range of branched α- limit dextrins. In addition, β-amylase (EC 3.2.1.2) catalyzes the hydrolysis and removal of successive maltose units from the non-reducing end of α - glucan chain. Alternatively, α- (1, 4) - glucosyl bonds may be cleaved phosphorolytically by starch phosphorylase to produce glucose-1-phosphate from successive glucosyl residues at the non-reducing end of the α - glucan chain.

Tetlow et al. [4] reported the way of process of starch degradation as depicted in Fig. 2.

2. STARCH PHOSPHORYLASES AND STARCH METABOLISM

Starch phosphorylase is an important enzyme for future studies and research. Starch phosphorylase (EC 2.4.1.1) catalyzes the reversible conversion of starch and inorganic phosphate into glucose-1-phosphate (G-1-P) and has been reported in many higher plants [5,6]. Glucose-1-phosphate can be used as an antibiotic in circulatory system therapy and in cardio-pathy therapy as it acts as a cytostatic compound [7].

The reaction can be described by the following equation:

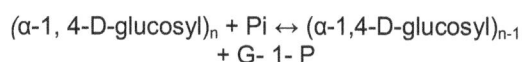

$$(\alpha\text{-}1, 4\text{-D-glucosyl})_n + P_i \leftrightarrow (\alpha\text{-}1,4\text{-D-glucosyl})_{n-1} + G\text{-}1\text{-}P$$

In plants, starch phosphorylase has been found in both cytosolic and chloroplastic forms. Starch phosphorylase has been found to be widely distributed in plant kingdom. It has been studied from pea seeds and leaves [8], potato tubers [9], maize [10], barley [11], germinating pea seeds [12], spinach [13], sweet potato [14], banana leaves [15], seaweed [16] and sweet corn [17]. It has also been studied from number of other plants. Role of starch phosphorylase has been described to be in starch degradation. The starch phosphorylase present in plastids is responsible for starch degradation due to the low glucose-1-phosphate and high Pi/g-1-p ratio found *in vivo* [1,18]. Contrarily, few scientists have described role of starch phosphorylase in primer synthesis which is used in starch synthase reaction [4,19-22]. Earlier, our laboratory illustrated the possible roles of phosphorylase in plants in starch metabolism. In plastid, starch granule may be attacked by glucan water dikinase, phosphoglucan water dikinase and α-amylase to form branched glucan and thereafter branched glucan is attacked by starch debranching enzyme to form linear glucan. The linear glucan is attacked by Pho1 to form glucose-1-phosphate which is either used in degradation route to form triose phosphate and triose phosphate comes out in the cytoplasm or linear glucan is acted upon by β-amylase to form maltose which is transported in the cytoplasm and there, it may be used in the formation of heteroglycan using disproportionating enzymes. The heteroglycan in the cytoplasm is degraded by Pho2 to form glucose-1-phosphate [23]. Rathore et al. [23] also described the role of Pho1 and Pho2 in the formation and degradation of glucose-1-phosphate in plastid and cytoplasm, respectively and also reported synthetic and degradative pathway/route in plastid for glucose-1-phosphate. In plastid, Pho 1 is responsible for glucose-1-phosphate formation from linear glucan as a substrate and this glucose-1-phosphate can be utilized in degradation route to form triose phosphate or can be utilized in synthetic route to form starch stored in starch granules with the help of starch synthase, starch debranching enzyme and ADP-glucose pyrophosphorylase enzymes. Rathore et al. [23] also proposed that maltose sugar in plastid can be produced from linear glucans by the action of β amylase.

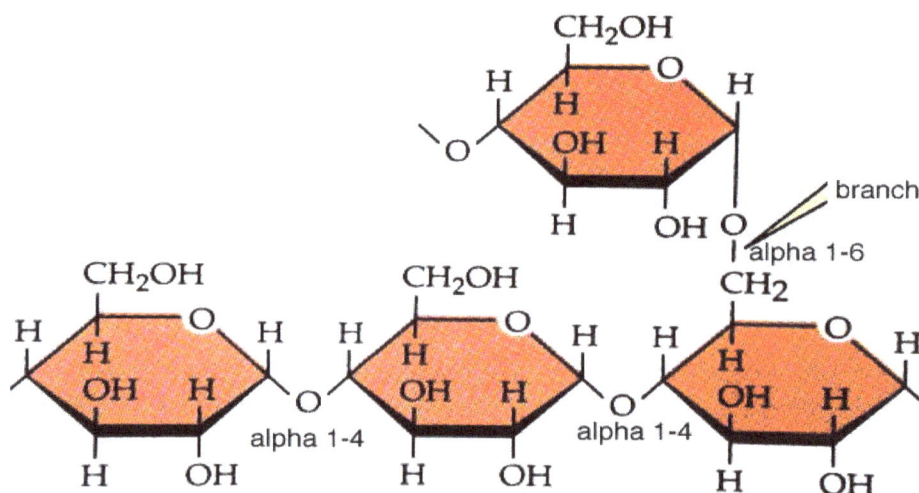

Fig. 1. Structure of starch (polymer of α-D- glucose)- Amylose, a constituent of starch has α-1,4 glycosidic bonds whereas, amylopectin, another constituent of starch has both α- 1, 4 glycosidic bonds and α- 1, 6 glycosidic bonds
(source:http://academic.brooklyn.cuny.edu/biology/bio4fv/page/starch.html)

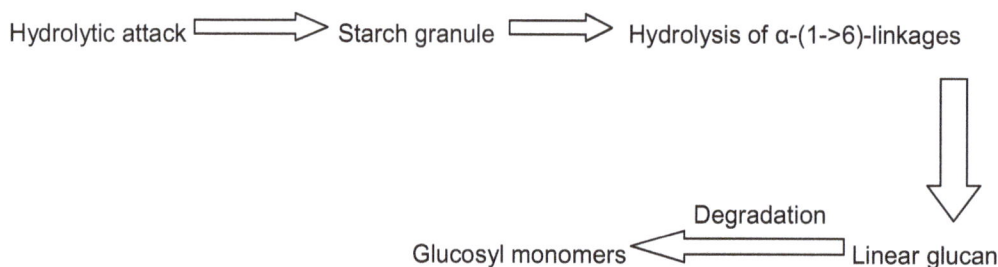

Fig. 2. The process of starch degradation in higher plants (Tetlow et al. [4])

Presence of multiple forms of starch phosphorylase has been reported in number of plant tissues. Post translational modifications viz. glycosylation, proteolytic modification and aggregation may be the reasons for multiplicity of starch phosphorylase [24]. In spite of multiple forms, reaction catalyzed by starch phosphorylase is same reflecting the characteristic feature of all starch degrading enzymes [23,25,26]. Synthesis of substrate starch for ADP glucose starch synthase has been favored by some multiple forms of starch phosphorylase [27]. In spinach leaves, multiple forms of pho1 in chloroplast and pho2 in cytoplasm have been reported and comparison has been done with respect to glycogen as substrate. As compared to Pho1, Pho2 has high affinity for glycogen [13].

Tsai and Nelson [10] reported different multiple forms pattern of starch phosphorylase during different developmental stages in maize endosperm and different multiple forms were assigned to have role in starch synthesis and degradation. Singh and Sanwal [28] reported presence of multiple forms in banana fruit pulp and showed role of different multiple forms in starch synthesis and degradation. Gerbrandy and Verleur [29] reported presence of multiple forms in growing and sprouting potato tubers and assigned role of different multiple forms in starch synthesis and degradation. However, it is considered that it has much limited role in degradation of starch *in vivo* since starches as well as starch phosphorylase both are impermeable to plastidial membrane. It is well known that starch synthesis occurs inside the plastid [23]. It has been shown that extra-plastidial enzymes are involved in novel glucan metabolism in the cytoplasm [30,31]. From so many studies, now it may be said that presence of multiple forms of starch phosphorylase is of physiological importance. Different multiple forms have different physiological functions. Intracellular localization of multiple forms also varies; a few are present in the cytoplasm and others in the plastids. It is well known that starch is not present in the cytoplasm and the multiple forms and starch phosphorylase as well as starch are not permeable through plastidial membrane. Also, starch is not known as a physiological substrate of cytoplasmic starch phosphorylase. More studies are required to unequivocally prove that heteroglycan present in the cytoplasm is the physiological substrate as has been reported by Fettke et al. [31].

Starch phosphorylase is regulated by pH, temperature, redox potential, oligosaccharides level and covalent modification [32,33]. Potato phosphorylase is stable at 30°C at pH 6 to 8 [34]. Lee and Kamogawa et al. [34,35] reported the inactivation of potato's starch phosphorylase at a temperature above 55°C.

3. STUDY OF STARCH PHOSPHORY-LASE IN DIFFERENT SOURCES

Mateyka and Schnarrenberger [36] compared starch phosphorylases of spinach leaves and corn leaves and reported that the two multiple forms of corn leaf phosphorylase are homologous to the spinach phosphorylase I and II. They reported that in corn phosphorylase, one multiple form having homology with spinach phosphorylase I is present in the cytoplasm of mesophyll cells whereas the other multiple form having homology with spinach phosphorylase II is present in the chloroplasts of bundle sheath cells. They reported that affinity of corn leaves starch phosphorylase and spinach starch phosphorylase I for phosphate and soluble starch is same, whereas affinity of corn leaves phosphorylase for glycogen is 15 times more compared to spinach phosphorylase [37,38].

Hsu et al. [39] purified and characterized the cytosolic form of starch phosphorylase from etiolated rice seedlings. Pho1 is inactivated by cyclodextrins and maltotetrose. However, thiol-reagents increased enzyme activity by 10-20% and metal ions viz. Zn^{+2}, Hg^{+2}, and Ag^+ reduced the enzyme activity. They also reported that there is no effect of sugar phosphate on Pho 1 enzyme activity in the absence of g- 1-p. Alignment was also done of partially sequenced enzyme purified from rice seedling with the peptide sequence of Pho 1 of *Oryza sativa* and *Triticum aestivum* and enzyme matched with four peptide sequences as shown in the Table 1

Witt and Sauter [40] reported partial purification of amylase and starch phosphorylase from poplar wood. They wanted to separate the multiple forms of starch degrading enzymes that were detected in extracts from poplar wood on native electrophoresis gels. They reported that endoamylase from poplar wood is not dependent on Ca^{2+} unlike other plant endoamylases. They also reported two multiple forms of starch phosphorylase from poplar wood, one present in the cytosol (type I) and it was retarded by immobilized amylopectin in acrylamide gels and it does not attack native starch granules, and the other form present in the amyloplast (type II) which was eluted at higher salt concentration from the anion exchanger and was hardly retarded on amylopectin containing gels. Weinhausel et al. [41] reported the isolation and characterization of starch phosphorylase from *Corynebacterium callunae*, a gram positive bacteria. This glucan phosphorylase depends on pyridoxal 5-phosphate for enzyme activity and has an important role in carbohydrate metabolism in bacteria. Glucan phosphorylase from *C. callunae* is induced by maltose. The enzyme from *C. callunae* is not catabolically repressed by D-glucose unlike glycogen phosphorylase from *Escherichia coli* which is controlled by catabolite repression. They also reported biochemical properties like effect of pH and temperature on the enzyme activity, cofactor analysis, effect of metabolites etc. They also determined the N-terminal sequence (Pro-Glu-Lys-Gln-Pro-Leu-Pro-Ala-Ala-Leu) and total amino acid composition of α- glucan phosphorylase. They reported that N-terminal region is entirely different compared to other glucan phosphorylases. However, total amino acid composition was similar to the enzyme isolated from *E. coli*, *Solanum tuberosum* and rabbit muscle. They explained the contribution of substrate chain length to binding energy by assuming two glucan binding sites, an oligosaccharide binding site composed of five subsites and a high affinity polysaccharide site separated from the active site. They gave a structural model of the molecular shape obtained with the help of small angle X-ray scattering. They found that model of *C. callunae* phosphorylase differs from rabbit muscle phosphorylase in size and axial dimension. They showed no inhibition of *C. callunae* phosphorylase by glucose, nucleotides, caffeine which distinguishes it from other regulated phosphorylases. All the α-glucan phosphorylases including glycogen phosphorylase from rabbit muscle, potato phosphorylase, *E. coli* glycogen phosphorylase have been reported to have pyridoxal 5'-phosphate (PLP) as a cofactor. Removal of PLP causes loss of the enzyme activity which may be restored on incubation of the enzyme with an excess of PLP. Reduction of the enzyme with sodium borohydride ($NaBH_4$) resulted in a material which is still 60% enzymatically active. However, all other PLP containing enzymes where PLP is involved in

catalysis are totally inactivated by reduction with $NaBH_4$. It binds as an imine through its C4 formyl group to an epsilon amino group of a lysine residue. There are reports where PLP has been shown to be involved in structural maintenance and/or acts as allosteric effector molecule [42]. Kumar and Sanwal [43-45] found absence of PLP in case of banana leaf and tapioca leaf phosphorylases. Kumar [42] reviewed role of PLP in phosphorylases in length.

Characterization of plastidial starch phosphorylase isolated from *Triticum aestivum* L. endosperm has been done by Tickle et al. [46]. They observed a single band of Pho activity using zymogram analysis in wheat endosperm crude extract and also at each stage of development and intensity of this band increased throughout the development stage until approximately to that 18 dpa. They studied the role of amyloplast localized starch phosphorylase isoform in starch synthesis in the developing wheat endosperm. Hukman et al. [47] showed that in endosperm, starch accumulation is highest between 10 to 30 dpa. Based on the results, they concluded that throughout this period, Pho 1 protein and activity in the endosperm support the role in synthesis of starch for Pho1. They presented the direct correlation between the expression of Pho 1 isoform in wheat endosperm and starch synthesis.

Kumar and Sanwal [48] reported that banana leaf phosphorylase amino acid composition is different from rabbit and potato phosphorylase, as they are poor in acidic and aromatic amino acids but enriched in basic and heterocyclic amino acids contents. Although starch phosphorylase is being studied since last more than 30 years, it has a special space in research area and now direction of research has changed from its purification and characterization to its immobilization, structural studies and its bioinformatics analysis. Precise methods for separation and purification of phosphorylases are required. Till date, people have concentrated on starch as a substrate for cytoplasmic phosphorylase despite the fact that starch is not present in the cytoplasm. Therefore, there is a strong need of characterization of the enzyme from a different angle.

Table 1. Matched amino acid sequences of Pho1 of *Oryza sativa* and *Triticum aestivum*

Significant hit	Species	Accession number	Match amino acid sequence
Pho1	*Oryza sativa*	gi|12025466	EGQEEIAEDWLEK TDQWTSNLDLLTGLR QLLNILGAVYR SGAFGTYDYAPLLDSLEGNSGFGR
Pho1	*Triticum aestivum*	Q9LKJ3	TDQWTSNL DLLTGLR

4. EXTRACTION OF STARCH PHOSPHORYLASE FROM DIFFERENT PLANT SOURCES

Purification of starch phosphorylase from young banana leaves was done long time [44]. The enzyme was extracted from young leaves by grinding in a waring blender and buffer employed was 0.01 M Tris-HCl, pH7.5 containing 0.02 M 2-mercaptoethanol and 0.05% Triton x-100. The enzyme has also been isolated from spinach leaves using 100 mM imidazole-HCl buffer, pH 7.0 containing 0.5% polyvinylpyrrolidone, 20 mm 2-mercaptoethanol and 0.1mM phenyl methyl sulfonyl fluoride (PMSF) [37]. The same buffer has been used for the extraction of the enzyme from corn (*Zea mays* L.) [36]. The HEPES-NaOH buffer (100 mM, pH 7.0) containing 10% glycerol, 1 mM EDTA, 1 mM dithioerythritol (DTE) and 0.1 mM phenyl methyl sulfonyl fluoride (PMSF) have been used for extraction of the enzyme from spinach [49]. The enzyme has been isolated from poplar wood by using 50 mM Tris-HCl buffer, pH 7.5, containing 10 mM $MgCl_2$ and 10 mM 2-mercaptoethanol [40]. The enzyme has also been isolated from etiolated rice seedlings by using 50 mM imidazole-HCl buffer, pH 7.0 having 1 mM PMSF, 1 mM benzamidine, 1 mM EDTA and 1 mM DTT [39]. Cabbage leaves starch phosphorylase has been extracted using 0.01 M Tris-HCl buffer, pH 7.5 containing 20 mM 2-mercaptoethanol and 50 mM EDTA [50]. The enzyme from wheat endosperm has been extracted using 100 mM Tris- HCl buffer, pH 7.5 [46]. The enzyme from potato tuber has been extracted using extraction buffer containing 100 mM HEPES buffer with 5 mM EDTA, 5 mM dithiothreitol, 0.2% (w/v) polyvinylpyrrolidone and 0.5% (w/v) bovine serum albumin [9]. Shatters and West [51] reported extraction of starch phosphorylase from the leaves of *Digitaria eriantha* using LCS buffer containing 0.33 mM sorbitol, 50 mM Tricine-NaOH, pH 8.0, 2 mM $NaNo_3$, 2 mM EDTA, 1 mM $MnCl_2$, 5 mM $MgCl_2$ and 5 mM K_2HPO_4. Srivastava et al. [52] reported extraction of starch phosphorylase from *Cuscuta reflexa* filaments using 0.05 M Tris – HCl, pH 7.6 containing 0.02 M 2-mercaptoethanol and 0.2% Triton X-100. Garg and Kumar [53] reported isolation of starch phosphorylase from seeds of Indian millet using 50 mM Tris – HCl buffer, pH 7.6 containing 10 mM 2- mercaptoethanol, 20 mM EDTA and 1% Triton –X -100. Kumar and Sanwal) [45] reported extraction of starch phosphorylase from tapioca leaves using 0.01 M Tris-HCl buffer , pH 7.5, 0.02 M 2- mercaptoethanol and 0.2% Triton X-100. Venkaiah and Kumar [54] reported extraction of the enzyme from sorghum leaves and also reported presence of multiple forms. They extracted the enzyme from leaves by using 0.05 M Tris HCl buffer, pH 7.6 containing 0.02 M EDTA, 0.02 M 2-mercaptoethanol, 1 % Trition-X-100 and 1 mM PMSF. Extraction of starch phosphorylase from germinating Bengal gram seeds has also been done using 50 mM Tris-HCl buffer, pH 7.6 containing 20 mM 2-mercaptoethanol, 10 mM EDTA and 1% Tritin-X-100 [55]. From the studies, it is clear that isolation medium of plant starch phosphorylases is more complex. In most of the cases, a non-ionic detergent, Triton-X-100 has been used which protects the enzyme from phenolics by dissociating protein quinone complex. Besides, it also ruptures the membranes of the organelles including that of plastids where one multiple form is localized. Depending on the stiffness of the tissue, grinding technique is selected. Most often and not, Waring blender is used, while some people also make use of pestle and mortar.

5. IMMOBILIZATION OF STARCH PHOSPHORYLASE

The term "Immobilized enzyme" refers to "enzyme physically confined or localized in a certain defined region of space with retention of catalytic activity and can be used repeatedly and continuously. The system of immobilization is consisted of the enzyme to be immobilized, the matrix for the support and the mode of attachment viz. physical adsorption, ionic linkages or through covalent bonding [56]. There are various methods of immobilization of enzymes summarized in Fig. 3.

An immobilized enzyme should have two essential functions namely catalytic function (CF) and non-catalytic function (NCF). Role of catalytic function is to drive the substrate conversion into the product within the given time and space and role of non-catalytic function is to make easy separation of the catalyst from the process, its reusability and process control [57]. Tischer and Kasche [58] reported some specific parameters of immobilized enzymes mentioned in Table 2.

Interaction of the enzyme and properties of carrier material control the characteristic features of the immobilized enzyme. After immobilization, measurement must be done for the yield of the enzyme activity, its efficiency, and operational stability in stirred tank fermenter and in number of cycles. With these data, one can determine the performance that is the product formed per unit enzyme or the amount of the enzyme required per unit product formed [59]. For industrial use of an immobilized enzyme, all the above mentioned parameters should be considered to make the process economically satisfied.

There are many advantages and disadvantages of immobilized enzymes over soluble enzymes (described in Table 3) [60].

Sheldon [61] described various techniques of enzyme immobilization. He critically reviewed various methods of immobilization especially in context of industrial processes particularly for use in non-aqueous media. He also emphasized on the use of novel supports viz. hydrogels, mesophorous silicas and smart polymers.

Immobilization of starch phosphorylase from different sources has been carried out on various matrices resulting in change of its physico-chemical properties. Starch phosphorylase from seeds of Indian Millet variety KB 560 has been immobilized on brick dust [53]. They reported the various characteristic features of immobilized enzyme with respect to soluble enzyme, like immobilized enzyme showed 3.5 fold purification compared to soluble enzyme, the temperature range was also different for both the enzymes. The optimum temperature for the enzyme activity has been found to be 37°C with two half temperature optima at 34°C and 40°C, while half life of immobilized enzyme was 9 hours at 30°C and 1 hour at 50°C. In case of soluble enzyme the half life was 5 hours at 30°C where as 30 minutes at 50°C. They also reported that the

enzyme can be reused many times and can be useful for glucose-1-phosphate production Kumar and Sanwal [62] reported immobilization of starch phosphorylase from mature banana leaves on polyacrylamide matrix. Immobilized enzyme retained about 80% activity and was more stable and could be reused several times. Garg and Kumar [50] immobilized starch phosphorylase from cabbage leaves for production of glucose -1-phosphate. They immobilized partially purified enzyme on hen egg shell and two fold purification was shown on immobilization compared to soluble enzyme and retention of enzyme activity upon immobilization was nearly 56%. Venkaiah and Kumar [63] immobilized starch phosphorylase from sorghum leaves using egg shell as solid support and showed retention of 46% enzyme activity. Srivastava et al. [64] reported immobilization of starch phosphorylase from *Cuscuta reflexa* on egg shell and reported retention of 50% enzyme activity upon immobilization. They reported that after ion exchange column chromatography, 95% pure glucose-1-phosphate was obtained as tested using phosphoglucomutase and glucose - 6-phosphate coupled enzyme assay. Conrath et al. [65] designed an enzyme electrode using immobilized acid phosphatase, glucose oxidase, mutarotase and phosphorylase with bovine serum albumin by glutaraldehyde cross linked dialysis membrane. It has been observed that after immobilization, there is increase in the specific activity of phosphorylase which is due to selective binding of phosphorylase and some other proteins on the solid supports. It is pertinent to mention that in most of the cases partially purified phosphorylase has been used for immobilization from the point of economics for production of glucose-1-phosphate. In most cases, there is increased optimum temperature and thermo-stability for the immobilized enzyme compared to soluble enzyme which may be exploited for industrial use. The increased thermo-stability and optimum temperature have been explained due to change in the conformation of the enzyme after immobilization.

There are number of other reports on immobilization of enzymes and microbial cells of commercial importance. Kourkoutas et al. [66] reviewed various immobilization technologies and matrices used in alcohol beverages. They described basic methods of cell immobilization and effect of immobilization on microbial cells. They also described various advantages of immobilized cells over free cell systems viz. prolonged activity and stability of biocatalyst, higher cell densities per unit bioreactor volume, increased substrate uptake and yield improvement, feasibility of continuous processing, increased tolerance to high substrate concentration and reduced end product inhibition, feasibility of low temperature fermentation leading to improved product quality, easier product recovery, regeneration and reuse of biocatalyst, reduced risk of microbial contamination, reduced maturation time for some products. Bakoyianis et al. [67,68] described various matrices used for immobilization of cells for wine making viz. inorganic supports namely mineral kissiris, a cheap, porous volcanic mineral found in Greece containing nearly 70% SiO_2, and organic supports namely alginates, cellulose, carrageenan, agar, pectic acid and chitosan.

Loukatos et al. [69] used porous g-alumina as support for cell immobilization for use in continuous wine making. Ogbonna et al. [70] used glass pellets covered with membrane of alginate as support for immobilization of *Saccharomyces cerevisiae* and *Schizosaccharomyces pombe* cells and used

immobilized cells for making wine. Otsuka [71] used cellulose covered with calcium alginate as support for immobilization of yeast and showed wine preparation using immobilized yeast. Lommi and Advenainen [72] used DEAE cellulose covered with an anion exchange resin for immobilization of yeast. Besides, enzymes have been immobilized on different solid supports for enhancing their activity and reusability.

Datta et al. [73] gave an overview of enzyme immobilization techniques and support materials used for immobilization. They mentioned that enzyme immobilization provides an excellent base for increasing availability of the enzyme to the substrate with greater turnover over a period of time. They assessed several natural and synthetic supports for their efficiency in enzyme immobilization. They reviewed natural supports viz. Alginate, chitosan and chitin, collagen, carrageenan, gelatine, cellulose, starch, pectin, sepharose, and synthetic polymers viz. Zeolites, ceramics, celite, silica, glass, activated carbon, charcoal used for enzyme immobilization.

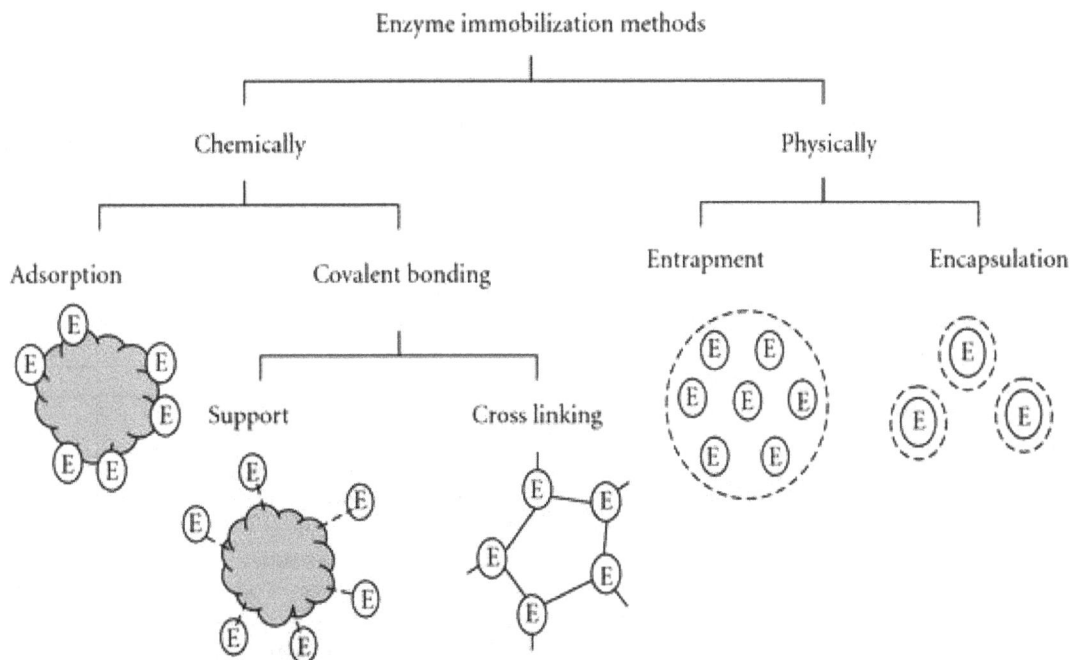

Fig. 3. Different methods of immobilization of the Enzymes
(http://www.hindawi.com/journals/er/2011/468292/Fig1/)

Table 2. Selected characteristic parameters of immobilized enzymes

Enzyme	Biochemical properties
	Molecular mass, prosthetic groups, functional groups on proteinsurface, purity (inactivating/protective function of impurities)
	Enzyme kinetic parameters
	Specific activity, pH-, temperature profiles, kinetic parameters for activity and inhibition, enzyme stability against pH, temperature, solvents, contaminants, impurities.
Carrier	**Chemical characteristics**
	Chemical basis and composition, functional groups, swelling behavior, accessible volume of matrix and pore size, chemical stability of carrier
	Mechanical properties
	Mean wet particle diameter, single particle compression behavior, flow resistance (for fixed bed application), sedimentation velocity (for fluidized bed), abrasion (for stirred tanks)
Immobilized enzyme	**Immobilization method**
	Bound protein, yield of active enzyme, intrinsic kinetic parameters (properties free of mass transfer effects)
	Mass transfer effects
	Consisting of partitioning (different concentrations of solutes inside and outside the catalyst particles), external and internal (porous) diffusion; this gives the effectiveness in relation to free enzyme. Determined under appropriate reaction conditions.
	Stability
	Operational stability (expressed as activity decay under working conditions), storage stability
	Performance
	Productivity (amount of formed product per unit or mass of enzyme) enzyme consumption (e.g. units kg–1 product, until half-)life

Table 3. Advantages and disadvantages of immobilized enzymes (Dicosimo et al. [60])

Advantages	Disadvantages
Amenable to continuous and batch formats	Loss of enzyme activity upon immobilization
Reuse over multiple cycles possible	Unfavorable alterations in kinetic properties
Improved stability over soluble enzyme forms	Cost of carrier and fixing agents
Favorable alterations in pH and temperature optima	Cost of immobilization process
Sequester enzyme from product stream	Mass transfer limitations
Co-immobilization with other enzymes possible	Subject to fouling

Chloroperoxidase, a heme-containing enzyme which exhibits catalase, peroxidase and cytochrome P450 enzyme activities in addition to halogenations reaction, has been covalently immobilized on to amino groups containing magnetic beads. For that, acrylate based magnetic beads were prepared by co-polymerization of monomers namely glycidylmethacrylate, methylmethacrylate and ethyleneglycol dimethacrylate via suspension polymerization. Thereafter, epoxy groups of the magnetic beads were converted into amino groups in the presence of ammonia during thermal precipitation iron oxide crystal in the beads' structures [74]. Kadima and Pickard [75] showed immobilization of chloroperoxidase on aminopropyl glass. Garg et al. [76] reported immobilization of urease on carboxyl terminated surface of glutathione capped gold nanoparticles and showed that immobilized enzyme can be exploited for biosensor fabrication, immunoassays and as in vivo diagnostic tools. Sahare et al. [77] reported immobilization of peroxidise enzyme onto porous silicon functionalized with 3-aminopropyldiethoxysilane and showed enhanced enzyme activity and stability of the enzyme upon immobilization. Nisha et al. [78] reviewed methods, applications and properties of immobilized enzymes. They reported that method of immobilization and solid support must be selected based on the type of application. They described various methods for enzyme immobilization and also discussed applications of immobilized enzymes in various industries viz. food industry, biodiesel industry, waste water treatment, textile industry, detergent industry etc. Yu et al. [79] reported non-covalent immobilization of cellulases using reversibly soluble polymers, Eudragit S-100 and Eudragit L-100 and showed use of immobilized enzyme in bio-polishing of cotton fabric. They also studied the characteristics of cellulose-Eudragit S-100 (CES) and cellulose Eudragit L-100 (CEL) using Fourier transform infrared (FTIR) spectra, circular dichroism (CD) spectra, and fluorescence spectra. The CES retained 51% enzyme activity whereas CEL retained only 42% enzyme activity after three cycles of repeated uses indicating that CES exhibits more stability compared to CEL. Endoglucanase has been immobilized using the technique of microencapsulation with gum Arabica as support, a natural polymer having biodegradable property. Encapsulation helps in retaining enzyme activity in the presence of detergents. Endoglucanase has been used in detergent and textile industries [80]. Kirby et al. [81] reported a simple method for high yield drug entrapment in liposomes using the technique of dehydration and rehydration. The same group reported that on encapsulation of cheese ripening enzymes in liposomes, there is acceleration of the rate of ripening of Cheddar cheese by nearly 100 fold over previously reported methods. They reported that objective measurement of cheese ripening by proteolytic indices, and subjective evaluation of flavor quality and intensity by trained taste panels indicated that cheeses are ripened by the microencapsulated enzyme in half the normal time thus making the process economical [82].

In food industry, enzyme inulinase is used for the production of fructose syrups. Fructose and fructo-oligosaccharides are important constituents in food, beverages and pharmaceutical industries due to their beneficial effects in diabetic patients, low carcinogenicity, in increasing iron absorption in children and higher sweetening property compared to sucrose. Besides, fructose also enhances flavour and color and product stability in food and bevarages

[83-85]. Yewale et al. [86] reported immobilization of inulinase from *Aspergillus niger* on chitosan and showed its usefulness in continuous inulin hydrolysis. Garlet et al. [87] immobilized inulinase non-covalently on multi-walled carbon nanotubes. They showed fast adsorption in about six minutes with a loading capacity of 51,047 units per gram support using 1.3% (v/v) inulinase concentration and 1: 460 adsorbent adsorbate ratio. They also reported 100% retention of immobilized enzyme activity during five weeks incubation at room temperature. Another example of carrier matrix for the support of enzyme immobilization is the amino functionalized $Fe_3O/SBA15$ composite used for *Candida rugosa* lipase, act as a biocatalyst for the interesterification of soyabean oil and methyl stearate [88].

6. PEPTIDE MAPPING

Peptide mapping called as peptide mass fingerprinting is a technique used to identify and characterize a purified protein. Here, protein is converted into peptide fragments either by chemical or enzymatic treatment and these fragments are separated and identified by comparing with the standard reference material. However, the known standard is not absolutely required if the protein is either unknown or uncharacterized previously. First two dimensional gel electrophoresis is carried out followed by the isolation and elution of the protein bands, which are subjected to treatment with a panel of proteases. The partial or complete digestion of the intended peptide generates number of fragments having different charge to mass ratio. The number of peptide fragments along with their respective size provides a peculiar pattern of peptide fragments which is compared with the peptide finger database of previously known proteins. The unknown protein may be characterized on the basis of similarities and dissimilarities with the earlier characterized proteins. Sometimes it is also used in conjunction with multidimensional HPLC for better resolution, if initial sample is not so pure. This method is preferable over the other methods available due to high sensitivity, specificity and versatility [89]. The magnitude of sensitivity is so much that even a single amino acid difference among peptides or proteins may be deduced using this technique. A specific map is developed through peptide mapping for a unique protein known as peptide map and it is the fingerprint of protein and is the end product of several chemical processes that provide a comprehensive understanding of the protein being analyzed. However, this technique may be simplified, if the protein of interest is previously well characterized as in our study on starch phosphorylase isoforms Pho1 and Pho2, which are well reported in number of other plants, but till date have not been studied using this technique.

There are four major steps in peptide mapping viz. isolation and purification of the protein, enzymatic or chemically selective cleavage of peptide bonds, separation of peptides by chromatography or electrophoresis and analysis and identification of peptides. Isolation and purification is much crucial step since contamination and impurities substantially hinder the resolution and consistent pattern of peptide mapping. It is the most tedious step because it needs lot of hit and trials. For starch phosphorylase, we have used two different approaches in our laboratory: (i) Classical chromatography techniques and (ii) Ultra filtration technique followed by native and SDS-PAGE electrophoresis. For selective cleavage of peptide bonds, there are two methods available viz. enzymatic method and chemical method. There are number of cleavage agents with their specificity of cleavage site(s). However, choice of cleavage agent depends upon the type of protein to be digested. For example, trypsin cleaves a peptide bond where carboxyl group has been contributed by either arginine or lysine, chymotrypsin cleaves a peptide bond where carboxyl group has been contributed by phenylalanine, tyrosine or tryptophan. The chemical agent, cyanogen bromide cleaves a peptide bond where carboxyl group has been contributed by methionine, whereas 2-nitro-5-thio-cyanobenzoic acid cleaves a peptide bond where amino group has been contributed by cysteine (http://www.nihs.go.jp/dbcb/Bio-Topic/peptide.pdf). Jeno et al. [90] reported a method where protein eluted in picogram quantity from Coomassie Blue stained two dimensional gel, may generate internal sequence data without blotting step. In the procedure, they reduced and alkylated the stained proteins with dithiothreitol and iodoacetamide in the presence of 1% sodium dodecyl sulfate and then digested with endoproteinase LysC.

Peptides generated after digestion of proteins from silver stained gels can be sequenced and analyzed by mass spectrometry. Shevchenko et al. [91] analyzed silver stained one dimensional gel of yeast protein by nano-electrospray tandem mass spectrometry which covered more than

1000 amino acids without any chemical modifications. They also mentioned that this silver staining procedure is preferable over Coomassie staining due to consumption of less time for sample preparation and also low level sequence analysis of proteins. Before digestion, number of parameters viz. pretreatment given to the sample and optimization of conditions like pH, temperature, time and amount of cleavage agent to be used must be taken into consideration (http://www.nihs.go.jp/dbcb/Bio-Topic/peptide.pdf).

For separation of peptides, RP-HPLC, ion exchange chromatography (IEC), capillary electrophoresis, and electrophoresis can be used. Among chromatographic separation techniques, HPLC and RP-HPLC are used very commonly. To use these techniques, selection of column, solvent, mobile phase, selection of the gradient, isocratic selection, temperature of the column, flow rate are very important. Intestinal calcium binding proteins after proteolytic digestion have been separated through HPLC and are considered as efficient techniques for the mixture of smaller peptides [92]. Cleveland et al. [93] reported peptide mapping by limited proteolysis and analysis by gel electrophoresis. After proteolytic digestion, peptides can be easily separated on 15% polyacrylamide gel having SDS due to large and different molecular weights of peptides. The generated band pattern is the characteristics feature of the protein with the particular proteolytic enzyme. Compared to different types of HPLC techniques, capillary electrophoresis technique has been proved to be better in recent few years. A very small size of sample say about 5 pico-mole quantity of protein can also be separated with capillary electrophoresis. This technique can separate proteins with high efficiency and in shorter analysis time [94]. Vaclav [95] reported improvement in the methodology of capillary electrophoresis including different separation modes viz. isoelectric focusing, isotachophoresis, affinity electrophoresis, multi-dimensional separations, zone electrophoresis, electro-chromatography.

On the basis of properties of peptides viz. difference in size, charge, hydrophobicity, shape and specific binding capacity, one can easily separate these peptide molecules through the above mentioned techniques. All the methods involving electro-migration depend on different properties of peptide molecule listed in Table 4.

6.1 Analysis and Identification of Peptides

After generation and separation of peptides, characterization of peptide fragments is done by Mass Spectroscopy (MS) either by inserting directly the peptide fragment or by online LC-MS for structure analysis. To determine the type of amino acid modification and to sequence a modified protein, tandem MS is used. To assign the disulphide bonds to various sulfhydryl containing peptides, a method has been developed by comparing the mass spectra of the digest before and after reduction. Major tools for the analysis and characterization of peptide finger prints or mapping are:

6.1.1 Mass spectroscopy

By fragmenting the sample in the instrument with multiple analyzers used in tandem mass spectrometer, structural information can be generated. Mass spectrometer has number of applications in biotechnology, pharmaco-genomics, clinical chemistry, environmental biology, geological analysis etc. due to abundant data of DNA sequencing and proteomics. Work has been enhanced to manage those data and with development in proteomics. After MS, protein directly can be compared with these databases and can be identified [96]. Ashcroft [97] reported number of ionization methods viz. electron impact (EI), electro spray ionization (ESI), matrix assisted laser desorption ionization (MALDI) etc. Among all the methods, MALDI and ESI are used widely for ionizing the bio-molecules. Karas et al. [98] reported that the use of matrix in MS yields ions without any fragmentation and this can be applicable to thermo- labile bio-molecules. Electro spray ionization has recently been upgraded with a powerful advancement of making complete ions of large and flimsy polar molecules which plays an important role in biological arrangement [99]. Gooley and Packer [100] reported that separation of proteins having high and low iso-electric points and membrane should be done by one dimensional gel electrophoresis instead of two dimensional gel electrophoresis.

6.1.2 MALDI- MS

In this technique, matrix absorbs the laser light as sample is embedded in it [101]. MALDI TOF is the commonly used technique which has reflector, post source decay capability delayed extraction. Here, matrix plays an important role. There are acidic and basic matrices according to their efficiency of generating mass spectrum. Selection of the matrix is important and depends upon the sample to be analyzed. For smaller sized proteins, basic matrix, 2-amino 4- methyl-5-nitro pyridine has been reported to be suited well and for oligonucleotides of short length and homo-polymers of thymidine, 2-amino-5-nitropyridine matrix is reported to give good results [102]. Gobom et al. [103] reported sample preparation and purification through nano-scale reverse phase column to improve the quality of spectrum of complex samples and contaminated mixtures by MALDI. They also reported that DHB (2,5 dihydroxy-benzoic acid) acts as more efficient matrix compared to CHCA (α-cyano 4 hydroxy cinnamic acid) which is generally used for fragile bio-molecules. It is necessary that peptide ions generated should be stable which is possible by using DHB.

In ESI, sample or analyte should be in solution form and ions from this solution are converted into gaseous phase for the mass spectrum analysis with the help of electrical energy and with increased sensitivity. ESI plays an important role in analyzing neutral compounds by converting them into ions by protonation or cationisation. The complete path of an ionic conversion of sample from solution to gaseous phase is a multistep procedure. Initially charge droplets in the form of fine spray are released and then evaporation of solvent takes place and finally from highly charged droplet, ion is ejected and enters into the triple quadrapole mass analyzer and then the spectrum generated by processing in data system or computer. From each peptide, ESI produces number of multiple charged ions. For online mass and sequence determination, triple quadrapole spectrometer with ion source is commonly used [104].

6.1.3 Protein databases

Protein databases play an important role in identifying the proteins from the peptide finger printing map of proteolytically digested protein sample. The availability of these databases exponentially enhances the use of proteomics that is the computational analysis of proteins. Due to the development in bioinformatics tools or homology search programs like BLAST and FASTA, it is possible to generate data of proteins which are modified or differ in the sequence from the database as these homology search programs display the sequences which are closely related to the sample peptide map [105]. Comparing the peptide map and computational data of proteins, one can study and screen the known proteins, detect translation errors, characterize post translational modification, locate disulphide bonds [106]. There are reports on the identification and characterization of microorganism proteins [107]. Dworzanski et al. [108] studied taxonomical classification and identified bacteria using mass spectrometric based proteomics. 'Profound' is an Bayseian algorithm based search engine which can be used for comparing the map with protein databases in order to identify proteins on the basis of their properties. On generation of low quality or if there is no sufficient data or if the sample contains mixture of proteins then also Profound identifies the protein reflecting the advantageous feature of the search engine [109]. Another database named as 'ProMEX' is a useful source for the study of plant system biology. It is a database for proteins and protein phosphorylation sites [110]. Brendt et al. [111] and Hummel et al. [110] reported an algorithm which depends upon number of parameters to identify true protein from the database. They also reported some possibilities to generate additional data because sometimes data collected by mass finger printing is not sufficient. To get more data, one can use another protease with different specificity or to improve spectrum quality, one should re-measure the spectrum with different parameters & to obtain important sequence information. For that, PSD MS/-MS peptide fragmentation technique can be used. Searching partial mass spectrometric peptide map of protein gives more specificity and error tolerance in mass spectrometric and database information. For map generated from four to five peptides, one can also use search engine like Protein Identification Resource (PIR) [112]. Yates [113] reported that development in computer algorithm is based on two different types of data generated, one from mass spectrometers and other from tandem mass spectrometer. It is possible to determine the molecular weight of

digested peptides to identify the protein through peptide mass finger printing using mass spectrometer and the databases. Another method is based on the information which is specific in fragmentation pattern and in sequence information generated from tandem mass spectrometer. This information is used in databases based on translated protein sequences and expresses tag sequences (EST) and it is very useful when the complete sequencing of genome of experimenting organism is not done but a large amount of expressed tag sequences are available. If regions of the primary structure are not clearly demonstrated by the peptide map, it might be necessary to develop a secondary peptide map. The goal of a validated method of characterization of a protein through peptide mapping is to reconcile and accounts for at least 95% of the theoretical composition of the protein structure. (http://www.nihs.go.jp/dbcb/Bio-Topic/peptide.pdf).

Peptide mapping or peptide fingerprinting is a plausible technique having huge potential for characterization of proteins previously unknown for their biochemical, biophysical and functional properties. However, much lesser literature is available on plant proteins especially on enzymes originated from plant sources. Yoon et al. [114] reported that botanical origin of commercial starches has been used for quality control of starch based products. Isolation of starch and its purification is considered quite cumbersome due to the fact that number of starch granules associated proteins (SGAPs) are available in plants such as soluble starch synthase I (SSI), starch branching enzyme II (SBEII), and granule-bound starch synthase I (GBSSI) which are irreversibly entrapped within the starch matrix and cannot be completely removed from starch granules by ordinary starch purification processes [115,116]. Cho et al. [117]

Table 4. Relationship between the properties and techniques for separation of peptide molecules

Properties of peptide molecule	Electro-migration Separation technique
Electrophoretic mobility (reflecting charge, size, and shape simultaneously.	Zone electrophoresis and isotachophoresis in a non sieving, free solution media.
Size	Zone electrophoresis in sieving media
Charge	Isoelectric .focusing, electrokinetic chromatography with ion – exchange pseudophase and by ion exchange electrochromatography
Hydrophobicity	Electrokinetic chromatography with micellar, microemulsion and br reverse phase chromatography.
Specific interaction with other biomolecules	Bio affinity electrophoresis.

reported identification of botanical origin of starch by using peptide mass fingerprinting of granule bound starch synthase. Starch is used as an ingredient in large number of products viz. Papers, foods, cosmetics, and drugs. This enzyme is an important enzyme of carbohydrate metabolism. Therefore, to identify any such protein including GBSSI is very important to conclude whether starch is originated from plant source. As these granules bound proteins or enzymes are the hallmark of starch originated from plant source, Cho et al. [117] emphasized that to analyze starch, number of methods such as microscopy, rheological measurements, X-ray diffractometry, FT-IR spectroscopy, and differential scanning calorimetry are available but for identifying the botanical origin of starches or starch based products, no reliable tool has been developed. However, for proteins, peptide mass fingerprinting (PMF) using MALDI-TOF-MS is one of the fastest and cheapest method.

Badgujar and Mahajan [118,119] reported peptide mass fingerprinting and N-terminal amino acid sequencing of cysteine protease named as Nivulian-II from *Euphorbia nivulia*. They also used peptide mass fingerprinting to elucidate details of the enzyme. They showed that this protease is a glycoprotein. Their mass fingerprinting analysis revealed peptide matches to Maturase K of *Banksia quersifolia*. The N-terminal sequence had partial homology with other cysteine proteases of biological origin and this work has very high impact in current trend of market where the requirement of cost effective proteases is increasing for various products such as innovative detergents. Yadav et al. [120] reported the characterization of a protein 'milin' from *Euphorbia milli,* a mollusc. Using the techniques of electrophoresis, gel filtration chromatography, mass spectrometry and other tests, they identified it as a homodimeric plant subtilisin like serine protease. The subunits of milin are differentially glycosylated affecting dimer association, solubility and proteolytic activity. The dimeric dissociation is SDS – insensitive and strongly temperature dependent but does not appear to be linked by disulphide bridges. They used peptide mass fingerprinting and *de novo* sequencing of the tryptic fragments but could not identify putative domains in the protein. The association studies of the milin suggested that it might play a role in the physiological function and in controlling Schistosomiasis (A mollusc originated disease). Singh and Dubey [121] studied a protease, procerain b from *Calotropis procera* and characterized it using peptide mass fingerprinting and N-terminal sequencing. This protease is shown to be involved in milk clotting and because of this property it may be exploited in cheese industry. It has also been mentioned that this protease may be a diet additive to stimulate digestion. N-terminal sequencing and PMF revealed that it is 80% similar to the Asclepain, which is a cysteine protease and functions like papain. Abdel Rahman et al. [122] analyzed this protein using electrospray ionization and matrix assisted laser desorption/ionization (MALDI) using hybrid quadruple time of flight tandem mass spectrometry. They showed presence of a signature peptide in the protein which may be useful for quantitative approach of the allergen prorein.

7. CONCLUSION

Starch in the form of insoluble granules accumulates in chloroplasts as the primary product of photosynthesis and acts as a primary energy source or reserve in higher plants. Starch phosphorylase (EC 2.4.1.1) is the key enzyme which catalyzes the reversible conversion of starch and inorganic phosphate into glucose-1-phosphate and has been reported in many higher plants. Multiple forms of the enzyme have been reported from many plant tissues which have been predicted to have different roles in starch metabolism. Here, various biochemical properties have been reviewed. Various techniques for enzyme immobilization have been discussed and reports on immobilization of starch phosphorylase from different sources and on different solid matrices have also been reviewed. Advantages of immobilized enzymes over soluble enzymes have been discussed. Disadvantages of immobilized enzymes viz. possible loss of enzyme activity upon immobilization, cost of carrier and fixing agent and immobilization process which affects the economics of the process have also been discussed. Peptide mapping technique for proteins and reports on peptide mapping of various proteins have also been illustrated in this review. In future research, there is need to elucidate the physiological role(s) of cytoplasmic phosphorylase (Pho 2). More research is required in exploiting the industrial uses of starch phosphorylase. Work should be carried out to search more useful immobilization techniques to prepare more thermo-stable enzyme which may be important for industrial purposes. Biopolymers can be prepared by exploiting starch phosphorylase which may be useful in food industry and can be of clinical importance. Besides, there is requirement of developing rapid and cheaper ways for purification of the enzyme. Research work must be carried out to construct chimeric starch phosphorylase having higher affinity for its substrates.

ACKNOWLEDGEMENTS

The authors acknowledge the facilities of the Department of Biotechnology, Ministry of Science and Technology, Government of India, New Delhi (DBT) under the Bioinformatics sub centre used in the preparation of the manuscript. The authors also acknowledge UGC, New Delhi for partial financial support.

COMPETING INTERESTS

Authors have declared that no competing interests exist.

REFERENCES

1. Newgard CB, Hwang PK, Fletterick RJ. The family of glycogen phosphorylases: Structure and function. Crit Rev Biochem Mol Biol. 1989;24:69-99.
2. Buchbinder JL, Rath VL, Fletterick RJ. Structural relationships among regulated and unregulated phosphorylases. Annu Rev Biophys Biomol Struct. 2001;30:191-209
3. Buleon A, Colonna P, Planchot V, Ball S. Starch granules: Structure and biosynthesis. Int. J. Biol. Macromol. 1998;23:85-112.
4. Tetlow IJ, Morell MK, Emes MJ. Recent developments in understanding the regulation of starch metabolism in higher plants. J. Exp. Bot. 2004b;55:2131-2145.
5. Fukui T. Plant Phosphorylases: Structure and Function. In: Akazawa T, Imaseki H (eds.), Martinus Nijhuff/ Dr. W. Junk Publishers, the Hague. N. Front. Plant Biochem. 1983;71-82.
6. Schachtele C, Steup M. α-1,4-Glucan phosphorylase from leaves of spinach (Spinacia oleracea L.). In situ localization by indirect immunofluorescence. Planta. 1986;167:444-451.
7. Weinhausel A, Nidetzky B, Rohrbach M, Blauensteiner B, Kulbe KD. A new maltodextrin-phosphorylase from Cornybacterium Callunae for the production of glucose 1- phosphate. Appl. Microbiol. Biotechnol. 1994;41:510-516.
8. Hanes CS. The breakdown and synthesis of starch by an enzyme system from pea seeds. Proc Roy Soc (London) B. 1940a;128:421–500.
9. Hanes CS. The reversible formation of starch from Glucose-1-phosphate catalyzed by potato phosphorylase. Proc R Soc (London) B. 1940b;129:174–208.
10. Tsai CS, Nelson OE. Two additional phosphorylases in developing maize seeds. Plant Physiol. 1969;44:159-167.
11. Baxter ED, Diffus CM. Enzymes of carbohydrate metabolism in developing Hordeum distichum grain. Phytochemistry. 1973;12:1923-1928.
12. Matheson NK, Richardson RH. Starch phosphorylase enzymes in developing and germinating pea seeds. Phytochemistry. 1976;15:887-892.
13. Steup M, Latzko E. Intracellular localization of phosphorylases in spinach and pea leaves. Planta. 1979;145:69-75.
14. Chang TC, Lee SC, Su JC. Sweet potato starch phosphorylase- purification and characterization. Agric Biol Chem. 1987;51:187-195.
15. Kumar A, Sanwal GG. Multiple forms of starch phosphorylase from banana leaves. Phytochemistry. 1977;16:327-328.
16. Yu S, Pederson M. Purification and properties of α-1,4- glucan phosphorylase from the red sea weed (Gracilaria sordida Harvey). Physiol Plant. 1991;81:149-155.
17. Lee EY, Braun JJ. Sweet corn phosphorylase:purification and properties. Arch. Biochem. Biophys. 1973;156:276–286.
18. Preiss, J, Levi C. Starch Biosynthesis and degradation. The biochem of plants. 1980;3:3371-3417.
19. Brisson N, Giroux H, Zollinger M, Camirand A, Simard C. Maturation and subcellular compartmentation of potato starch phosphorylase. Plant Cell. 1989;1:559-566.
20. Albrecht T, Koch A, Lode A, Greve B, Schneider-Mergener J, Steup M. Plastidic (Pho1- type) phosphorylase isoforms in potato (Solanum tuberosum L.) plants: expression analysis and immunochemical characterization. Planta. 2001;213:602-613.
21. Dauvillée D, Chochois V, Steup M, Haebel S, Eckermann N, Ritte G, Ral JP, Colleoni C, Hicks G, Wattebled F, Deschamps P, d'Hulst C, Liénard L, Cournac L, Putaux JL, Dupeyre D, Ball SG. Plastidial

phosphorylase is required for normal starch synthesis in *Chlamydomonas reinhardtii*. Plant J. 2006;48:274-285.

22. Satoh H, Shibahara K, Tokunaga T, Nishi A, Tasaki M, Hwang SK, Okita TW, Kaneko N, Fujita N, Yoshida M, Hosaka Y, Sato A, Utsumi Y, Ohdan T, Nakamura Y. Mutation of the plastidial. α -glucan phosphorylase gene in rice affects the synthesis and structure of starch in the endosperm. Plant Cell. 2008;20:1833-1849.

23. Rathore RS, Garg N, Garg S, Kumar A. Starch phosphorylase: Role in starch metabolism and biotechnological applications. Crit Rev Biotechnol. 2009;29:214-224.

24. Yu S, Pederson M. One step purification to homogeneity and isoforms of α -1,4-glucan phosphorylase of red sea weeds. Plant Physiol. Biochem. 1991a;29:9-16.

25. Smith AM, Zeeman SC, Thorneycroft D, Smith SM. Starch mobilization in leaves. J Exp. Bot. 2003;54:577-583.

26. Li Z, Morell MK, Rahman S. Rice and products thereof having starch with an increased proportion of amylase. US patent application 20070300319; 2007.

27. Frydman RB, Slabnik E. The role of phosphorylase in starch biosynthesis. Ann. New York Acad. Sci. 1973;210:153-169.

28. Singh S, Sanwal GG. Multiple forms of α-glucan phosphorylase in banana fruits. Phytochemistry. 1976;15:1447-1451.

29. Gerbrandy SJ, Verleur JD. Phosphorylase isoenzymes: Localization and occurrence in different plant organs in relation to starch metabolism. Phytochemistry. 1971;10:261-266.

30. Fettke J, Poeste S, Eckermann N, Tiessen A, Pauly M, Geigenberger P, Steup M. Analysis of cytosolic heteroglycans from leaves of transgenic potato (*Solanum tuberosum L.*) plants that under- or overexpresses the Pho 2 phosphorylase isozyme. Plant Cell Physiol. 2005;46:1987-2004.

31. Fettke J, Eckermann N, Kötting O, Ritte G, Steup M. Novel starch-related enzymes and carbohydrates. Cell. Mol. Biol. 2007;52:OL883–OL904.

32. Smith AM, Zeeman SC, Smith SM. Starch degradation. Annu. Rev. Plant Biol. 2005;56:73-98.

33. Orzechowski S. Starch metabolism in leaves. Acta Biochem Polonica. 2008;55:435-445.

34. Lee YP. Potato phosphorylase. I. Purification, physico-chemical properties and catalytic properties. Biochem. Biophys. Acta. 1960;43:18-24.

35. Kamogawa A, Fukui T, Nikuni A. Potato α-glucan phosphorylase: Crystallization, amino acid composition and enzymatic reaction in the absence of primer. J. Biochem. 1968;63:361-369.

36. Mateyka C, Schnarrenberger C. Purification and Properties of Mesophyll and Bundle Sheath Cell α-Glucan Phosphorylases from *Zea mays* L. Equivalence of the Enzymes with the Cytosol and Plastid Phosphorylases from Spinach. Plant Physiol. 1988;86:417-422.

37. Preiss J, Okita TW, Greenberg E. Characterization of the spinach leaf phosphorylases. Plant Physiol. 1980;66:864-869.

38. Steup M, Schachtele C, Latzko E. Purification of a non-chloroplastic α- glucan phosphorylase from spinach leaves. Planta. 1980;48:168-173.

39. Hsu JH, Yang CC, Su JC, Lee PD. Purification and characterization of a cytosolic starch phosphorylase from etiolated rice seedlings. Bot. Bull. Acad. Sin. 2004;45:187-196.

40. Witt W, Sauter JJ. Partial Purification of Amylases and Starch Phosphorylases from Poplar Wood. Plant Physiol. 1994;146:15-21.

41. Weinhausel A, Griessler R, Krebs A, Zipper P, Haltrich D, Kulbe KD, Nidetzky B. α -1,4-D-glucan phosphorylase of gram-positive Corynebacterium callunae: Isolation, biochemical properties and molecular shape of the enzyme from solution X-ray scattering. Biochem J. 1997;326:773-783.

42. Kumar A. Starch phosphorylase in plants. J. Sci. Ind. Res. 1989;48:568-576.

43. Kumar A, Sanwal GG. Characterization of purified starch phosphorylase from mature banana (*Musa paradisiacal*) leaves. Indian J. Biochem. Biophys. 1981;18:421-424.

44. Kumar A, Sanwal GG. Purification and physicochemical. properties of starch phosphorylase from young banana leaves. Biochemistry. 1982a;21:4152-4159.

45. Kumar A, Sanwal GG. Starch phosphorylase from tapioca leaves: Absence of pyridoxal phosphate. Arch Biochem Biophys.1982b;217:341-350.

46. Tickle P, Burrell MM, Coates SA, Emes MJ, Tetlow IJ, Bowsher CG.

Charecterization of plastidial starch phosphorylase in *Tritcium aestivum* L. endosperm. J. Plant Physiol. 2009;166:1465-1478.

47. Hurkman WJ, McCue KF, Altenbach SB, Korn A, Tanaka CK, Kothari KM, Johnson EL, Bechtel DB, Wilson JD, Anderson OD, DuPont FM. Effect of temperature on expression of genes encoding enzymes for starch biosynthesis in developing wheat endosperm. Plant Science. 2003;164:873-881.

48. Kumar A, Sanwal GG. Amino acid composition of starch phosphorylase from banana (*Musa paradisiaca*) leaves. Ind. J. Biochem. Biophys. 1982d;19:62-63.

49. Hammond JB, Preiss J. Spinach Leaf Intra and Extra Chloroplast Phosphorylase Activities during Growth. Plant Physiol.1983;73:709–712.

50. Garg N, Kumar A. Immobilization of starch phosphorylase from cabbage leaves: Production of glucose-1-phosphate. Braz J Chem Eng. 2008;25:229-235.

51. Shatters RG Jr, West SH. Purification and characterization of non-chloroplastic α-1, 4-glucan phosphorylases from leaves of Digitaria eriantha Stent. J. of plant physiol. 1996;149:501-509.

52. Srivastava S, Nighojkar A, Kumar A. Purification and chatrecterization of starch phosphorylase from *Cuscuta reflexa* filaments. Phytochemistry. 1995;39:1001-1005.

53. Garg S, Kumar A. Immobilization of starch phosphorylase from seeds of Indian millet (*Pennisetum typhoides*) variety KB560. Afr J Biotechnol. 2007;6:2715-2720.

54. Venkaiah B, Kumar A. Multiple forms of starch phosphorylase from Sorghum leaves. Phytochemistry. 1996;41:713-717.

55. Upadhye SP, Kumar A. Immobilization of starch phosphorylase from Bengal gram seeds: Production of glucose-1-phosphate. The Genet Eng Biotechnol. 1996;16:145-151.

56. Brena BM, Viera FB. Immobilization of enzyme and cells. Methods in Biotec. 2006;22:15-30.

57. Cao L. Carrier-bound Immobilized Enzymes: Principles, Application and Design. Wiley-VCH Verlag GmbH & Co; 2006. ISBN:9783527607662. DOI:10.1002/3527607668.

58. Tischer W, Kasche V. Immobilized enzymes: Crystals or carriers? Trends Biotechnol. 1999;17(8):326-335.

59. Tischer W, Frank Wedekind. Immobilized Enzymes: Methods and Applications. Top. curr. chem. 1999;200:96-118.

60. DiCosimo R, McAuliffe J, Poulose AJ, Bohlmann G. Industrial use of immobilized enzymes. Chem. Soc. Rev. 2013;42:6437-6474.

61. Sheldon RA. Enzyme immobilization: The quest for optimum performance. Adv. Synth. Catal. 2007;349:1289-1307.

62. Kumar A, Sanwal GG. Immobilized starch phosphorylase from mature banana (*Musa paradisiaca*) leaves. Indian J. Biochem. Biophysics.1981;114-119.

63. Venkaiah B, Kumar A. Egg shell bound starch phosphorylase packed bed reactor for the continuous production of glucose-1-phosphate. J. Biotechnol. 1994;36:11-17.

64. Srivastava S, Nighojkar A, Kumar A. Immobilization of *Cuscuta reflexa* starch phosphorylase: Production of glucose-1-phosphate using bioreactors. J Ferment Bioeng. 1996;81:355-357.

65. Conrath N, Griindig B, Hiiwel St, Cammann K. A novel enzyme sensor for the determination of inorganic phosphate. Anal. Chim. Acta. 1995;309:47-52.

66. Kourkoutas Y, Bekatorou A, Banat IM, Marchant R, Koutinas AA. Immobilization technologies and support materials suitable in alcohol beverages production. Food Microbiol. 2004;21:377-397.

67. Bakoyianis V, Kanellaki M, Kalliafas A, Koutinas AA. Low temperature wine-making by immobilized cells on mineral kissiris. J. Agric. Food Chem.1992;40: 1293–1296.

68. Bakoyianis V, Kana K, Kalliafas A, Koutinas AA. Low temperature continuous wine-making by kissiris supported biocatalyst. J. Agric Food Chem. 1993; 41:465–468.

69. Loukato P, Kiaris M, Ligas I, Bourgos G, Kanellaki M, Komaitis M, Koutinas AA. Continuous wine-making by g-alumina-supported biocatalyst. Quality of the wine and distillates. Appl. Biochem. Biotechnol. 2000;89:1–13.

70. Ogbonna JC, Amano Y, Nakamura K, Yokotsuka K, Shimazu Y, Watanabe M, Hara SA. Multistage bioreactor with replaceable bioplates for continuous wine fermentation. Am. J. Enol. Vitic.1989;40: 292–297.

71. Otsuka K. Wine making. Jpn. Kokkai Tokkyo Koho. 80, 159, 789. Tita, O, Jascanu V, Tita M, Sand C. Economical

comparative analysis of different bottle fermentation methods. Proceedings of the AVA, International Conference on Agricultural Economics, Rural Development and Informatics in the New Millennium, Hungary. 2003;1980.

72. Lommi H, Advenainen J. Method using immobilized yeast to produce ethanol and alcoholic beverages. European Patent Application EP. 1990;361:165.

73. Datta S, Christena LR, Rajaram YRS. Enzyme immobilization: an overview on techniques and support materials. Biotech. 2013;3:1-9.

74. Bayramoğlu G, Kirlap S, Yilmaz M, Topaare L, Arıca MY. Covalent immobilization of chloroperoxidase onto magnetic beads: Catalytic properties and stability. Biochem. Eng. J. 2008;38:180–188.

75. Kadima TA, Pickard MA. Immobilization of chloroperoxidase on aminopropyl-glass. Appl. Environ. Microbiol. 1990;56:3473–3477.

76. Garg S, De A, Mozumdar S. pH dependent immobilization of Urease on glutathione capped gold nanoparticles. J Biomed Material Res. Epub ahead of print; 2014.

77. Sahare P, Ayala M, Duhalt RV, Vivechana A. Immobilization of peroxidase enzyme onto theporous silicon structure for enhancing its activity and stability. Nanoscale Research Letters. 2041;9:409.

78. Nisha S, Arun K S, Gobi N. A Review on Methods, Application and Properties of Immobilized Enzyme. Chem. Sci. Rev. Lett. 2012;1:148-155.

79. Yu Y, Yuan J, Wang Q, Fan X, Wang P, Cui L. Non-covalent immobilization of cellulases using the reversibly soluble polymers for bio-polishing of cotton fabric. Biotechnol. Appl. Biochem; 2014. DOI: 10.1002/bab.1289. [Epub ahead of print].

80. Ramakrishnan A, Pandit N, Badgujar M, Bhaskar C, Rao M. Encapsulation of endoglucanase using a biopolymer Gum Arabic for its controlled release. Bioresour. Technol. 2007;98:368-372.

81. Kirby CJ, Gregoriadis G. Dehydration – rehydreation vesicles: Simple method for high yield drug entrapment in lisosomes. Biotechnol. 1984;2:979.

82. Kirby CJ, Brooker BE, Law BA. Accelerated ripening of cheese using liposome encapsulated enzyme. Int J of Food Sci and Tech. 1987;22:355-375.

83. Roy I, Sardar M, Gupta MN. Exploiting unusual affinity of usual polysaccharides for bioseparation of enzymes on fluidized beds. Enzyme. Microb. Technol. 2000;27:53-65.

84. Hanover LM, White JS. Manufacturing, compositon, and aplications of fructose. Am. J. Clin. Nutr. 1993;58:724S-732S.

85. Treichel H, Mazutti M, Maugeri Filho F, Rodrigues MI. Technical viability of the production, partial purification and characterisation of inulinase using pretreated agroindustrial residues. Bioproc Biosyst Eng. 2009;32:425-433.

86. Yewale T, Singhal RS, Vaidja AA. Immobilization of inulinase from Aspergillus niger NCIM 945 on chitosan and its application in continuous inulin hydrolysis. Biocatal Agric Biotechnol. 2013;2:96–101.

87. Garlet TB, Weber CT, Klaic R, Foletto EL, Jahn SL, Mazutti MA, Kuhn RC. Carbon Nanotubes as Supports for Inulinase Immobilization. Molecules. 2014;19:14615-14624.

88. Zang X, Xie W. Enzymatic Interesterification of Soybean Oil and Methyl Stearate Blends Using Lipase Immobilized on Magnetic Fe3O4/SBA-15 Composites as a Biocatalyst. Oleo Sci. [Epub ahead of print]; 2014.

89. Primrose S, Twyman R. In Principles of gene manipulation and genomics. Blackwell Publishing press; 2006.

90. Jenö P, Mini T, Moes S, Hintermann E, Horst M. Internal sequences from proteins digested in polyacrylamide gels. Anal.Biochem. 1995;224:75-82.

91. Shevchenko A, Wilm M, Vorm O, Mann M. Mass spectrometric sequencing of proteins silver stained polyacrylamide gels. Anal. Chem. 1996;68:850-858.

92. Fullmert CS, Wasserman RH. Analytical Peptide Mapping by High Performance Liquid Chromatography .APPLICATION TO Intestinal calcium binding proteins. J. Biol. Chem. 1979;254:7208-7721.

93. Cleveland DW, Fischer SG, Kirschner MW, Laemmli UK. Peptide Mapping by Limited Proteolysis in Sodium Dodecyl Sulfate and Analysis by Gel Electrophoresis. J. Biol. Chem. 1977;252:1102-1106.

94. Cobb KA, Novotny MV. Peptide mapping of complex proteins at the low -picomole

level with capillary electrophoretic separations. Anal. chem. 1997;64:879-886.

95. Vaclav K. Recent advances in capillary electrophoresis of peptides Electrophoresis. 2001;22:4139-4162.

96. Yates JR. 3rd. Mass spectrometry from genomics to proteomics. Trends Genet. 2000;16:5-8.

97. Ashcroft AE. In Ionization Methods in Organic Mass Spectrometry. The Royal Society of Chemical Sciences press, London. 1997;1-26.

98. Karas M, Bachmann D, Bahr U, Hillenkamp F. Matrix-assisted ultraviolet laser desorption of non – volatile compounds. Int. J. mass spectrom. 1987;78:53-68.

99. Fenn JB, Mann M, Meng CK, Wong SF, Whitehouse CM. Electrospray ionization for mass spectrometry of large biomolecules. Science. 1989;24:64-71.

100. Gooley AA, Packer NH. The importance of co- and post-translational modifications in proteome projects. Proteome Research: New Front, Functional Genomics Principles and Practice. Wilkins W, Wilhams K, Appei R. Hochstrasser D. Springer. 1997;65-91.

101. Rusconi F, Schmitter JA. Sample preparation for analysis by MALDI-TOF mass spectrometry. Analysis. 1998;26:M13-M17.

102. Fitzgerald MC, Parr GR, Smith LM. Basic matrices for the matrix-assisted laser desorption/ionization mass spectrometry of proteins and oligonucleotides. Anal. Chem. 1993;65:3204-3211.

103. Gobom J, Nordhoff E, Migorodskaya E, Ekman R, Roepstorff P. Sample purification and preparation techniques based on nano scale reversed –phase columns for the sensitive analysis of complex peptide mixtures by matrix assisted laser desorption/ ionization mass spectrometry. J. Mass Spectrom. 1999;34:105-116.

104. Lahm HW, Langen H. Mass spectrometry: A tool for the identification of proteins separated by gels. Electrophoresis. 2000;21:2105-2114.

105. Johnson RS, Davis MT, Taylor JA, Patterson SD. Informatics for protein identification by mass spectrometry. Mass spectrometry in proteomics. 2005;35:223-236.

106. Pappin DJ, Horjup P, Bleasby AJ. Rapid identification of proteins by peptide – mass fingerprinting. Curr. Biol. 1993;3:327-32.

107. Demirev PA, Fenselau C. Mass spectrometry for rapid characterization of microorganisms. Annu. Rev. Anal. Chem. 2008;1:71-93.

108. Dworzanski JP, Snyder AP. Classifiation and identification of bacteria using mass spectrometry-based proteomics. Expert Rev Proteomics. 2005;2:863-878.

109. Zhang W, Chait BT. Profound: An expert system for protein identification using mass spectrometric peptide mapping information. Anal. Chem. 2000;72:2482-2489.

110. Hummel J, Niemann M, Wienkoop S, Schulze W, Steinhausr D, Selbig J, Wather D, Weckwerth W. Pro MEX: A mass spectral reference databases for proteins and protein phosphorylation sites. BMC. 2007;8:216.

111. Brendt P, Hobohm U, Langen H. Reliable automatic protein identification from matrix-assisted laser desorption/ionization mass spectrometric peptide fingerprints. Electrophoresis. 1999;20:3521-3526.

112. Mann M, Hojrup P, Roepsorff P. Use of mass spectrometric molecular weight information to identify protein sequence databases. Biol. Mass Spectrom. 1993;22:338-345.

113. Yates JR. 3rd. Databases searching using mass spectrometry data. Electrophoresis. 1998;19(6):893-890.

114. Yoon JW, Jung JY, Chung HJ, Kim MR, Kim CW, Lim ST. Identification of botanical origin of starches by SDS –PAGE analysis of starch granule-associated proteins. J of cereal Sci. 2010;52:321-326.

115. Koziol AG, Marquez BK, Huebsch MP, Smith JC, Altosaar I. The starch granule associated proteomes of commercially purified starch reference materials from rice and maize. J proteomics. 2012;75:993-1003.

116. Mu-Forster C, Huang R, Powers JR, Harriman RW, Knight M, Singletary GW, Keeling PL, Wasserman BP. Physcial association of starch biosynthetic enzymes with starch granules of maize endosperm. Plant Physiol. 1996;111:821-829.

117. Cho DH, Han JA, Lim ST. Identification of botanical origin of starch by using peptide mass fingerprinting of granule- bound starch synthase. J Cereal Sci. 2013;57:264-266.

118. Badgujar SB, Mahajan RT. Peptide Mass Fingerprinting and N-Terminal Amino Acid equencing of Glycosylated Cysteine Protease of *Euphirbia nivulia* Buch.-Ham. J Amino Acids. 2013;1-6.

119. Badgujar SB, Mahajan RT. Charecterization of Thermo- and Detergent Stable Antigenic Glycosylated Cysteine Proteases of *Euphorbia nivulia* Buch.-Ham. and Evalution of Its Ecofriendly Applications. Scientific World J. 2013;1-12

120. Yadav SC, Jagannadham MV, Kundu S, Jagannadham MV. A kinetically stable plant substilase with unique peptide mass fingerprints and dimerization properties. Biophys. Chem. 2009;139:13-23.

121. Singh AN, Dubey VK. Exploring applications of procerain b, a novel protease from Calotropis procera, and characterization by N-terminal sequencing as well as peptide mass fingerprinting. Appl. Biochem. Biotechnol. 2011;164:573-80.

122. Abdel Rahman AM, Lopata AL, O'Hehir RE, Robinson JJ, Banoub JH, Helleur RJ. Charecterization and de novo sequencing of snow crab tropomyosin enzymatic peptides by both electrospray ionization and matrix-assisted laser desorption ionization QqToF tandem mass spectrometery. J. Mass Spectrom. 2010;45:372-81.

7

Biochemical Characterization, 16S rRNA Sequence Analysis and Multiple Sequence Alignment of Bacteria Isolated from Fermented and Unfermented Coconut

Hailu Weldekiros Hailu[1,2*], Viktor Abdallah[1], Ari Susilowati[1], Saleh Muhammed Raqib[1,3] and Ali Ramadan Ali ALatawi[1,4]

[1]*Department of Biology, Bioscience postgraduate program, Faculty of Mathematics and Natural Sciences, Sebelas Maret University, Surakarta, Indonesia.*
[2]*Department of Biology, College of Science, Bahir Dar University, Bahir Dar, Amhara, Ethiopia.*
[3]*Department of Biotechnology, Faculty of Agriculture, Bangladesh Agriculture University, Mymensingh, Bangladesh.*
[4]*Department of Chemistry, Faculty of Science, Sirte University, Sirte, Libya.*

Author contributions

This research was carried out in collaboration between all authors. All authors designed and conducted the study. Author HWH managed literature search, performed biochemical, molecular and bioinformatics analysis and compiled the manuscript. Author AS supervised the research. All authors read and approved the final manuscript

Editor(s):
(1) Chung-Jen Chiang, Department of medical laboratory Science and Biotechnology, China Medical University, Taiwan.
Reviewers:
(1) Anonymous, Japan.
(2) Satpal Singh Bisht, Dept. of Zoology, Kumaun University, India.

ABSTRACT

Coconut (*Cocos nucifera L.*) is a tropical and subtropical plant which has great versatility due to many benefits of its different parts. Appropriate methods such as biochemical, morphological, serological, physiological and molecular techniques are required to clearly differentiate beneficial from harmful bacteria that are found in coconut. The objectives of this research were to characterize bacterial diversity in coconut based on biochemical tests and molecular techniques, to study the evolutionary relationship among the bacteria isolates based on 16S rRNA BLAST

Corresponding author: Email: hawelbirtu@hotmail.com

analysis and Multiple Sequence Alignment. Eight bacteria isolates, 3 from fermented and 5 from non-fermented coconut were identified. Biochemical characterization, polymerase chain reaction amplification of 16S rRNA using universal primers (63f-1387r), BLAST analysis and multiple sequence analysis were performed. The result indicated that the bacteria were detected as *Klebsiella pneumonia, Enterobacter aggromerans, Pseudomonas, Ralstonia pickettii* and *Burkholderia*. The first two were carbohydrate fermenting while the last three were non-fermenting bacteria. All these species are related to human pathogenicity. This can be related to hygiene in food process and handling procedures. Therefore, to favour the selective growth of beneficial bacteria, appropriate environmental hygiene is required.

Keywords: 16S rRNA; Bacteria; Biochemical; BLAST; Multiple sequence analysis; PCR.

1. INTRODUCTION

Coconut (*Cocos nucifera* L.) has been envisaged as a valuable source of several commercial products for human life [1], as well as the most important and hugely grown palm tree worldwide [2]. Each sections of the plant is useful and, in many cases, human is dependent on it [3]. The leaf and trunk supply building material, and the root is utilized as medicine [4]. The fruit is the most demandable part for food and sale. The envelope called mesocarp is processed into rope, carpets, geotextiles and growing media. The hard brown shell (endocarp) can be processed into very high-quality activated charcoal. The inner part of the nut (endosperm) is divided into two edible parts: a white kernel and a clear liquid (coconut water). Moreover, Coconut oil is consumed in tropical countries for many centuries. Studies done on native diets high in coconut oil expenditure reveal that the population is usually in good health. Coconut oil carries a long shelf life and is capitalized in baking industries, processed foods, infant formulae, pharmaceuticals, cosmetics and as hair oil [5].

Coconut meat can be used either as fermented or unfermented form. The protein content, crude fat and crude fiber of fermented defatted coconut increased while the carbohydrate, moisture and ash content decreased as the fermentation period increased [6].

Most mercantile grade coconut oils are derived from copra, the dried kernel of coconut. It can be produced by: smoke drying, sun drying, kiln drying, or derivatives or a merger of these three. *Nevertheless, virgin coconut oil* has more health benefits as compared to coconut oil extracted from copra and it is very durable with a shelf life of several years [5,7].

Coconut milk raw material used for virgin oil production mainly consists of (w/w) 21.3% fat, 2.0% protein, 2.8% carbohydrate and 2.1% of sugar together with some vitamins and trace elements. Virgin Coconut Oil (VCO) consists of C_8, C_{10}, C_{12} and C_{14} fatty acid with lauryl acid has the most composition of 47-53% among all saturated fatty acid [8,9]. Around 45 to 50% of fatty acids of coconut oil form lauric acid. This acid is cognizant to kill viruses, bacteria, yeast, and fungi etc. that are enveloped in a phospholipid membrane. [5,10].

Some microorganisms however, have adapted with the environments circumstances and could actively break the compounds in coconut milk and emerged as either beneficial or harmful. Additionally, disintegration of the linkage of the protein, fat and carbohydrate could be conducted by the inoculation of effective microbes having proteolysis and amyloid enzymes on to the freshly extracted coconut milk to coagulate the protein and further centrifugation can separateprotein from fat, carbohydrate water. With this method, lauryl acid will not be decomposed [9,11].

Under the non-sanitary environmental conditions low quality Virgin Coconut Oil is produced. The wild organism may grow producing several compounds, which affects the quality of the oil. Therefore, prior and proper identification of microorganism is important to avoid microbial associated risks. Some bacteria can be identified by direct microscopic observation. However, definitive identification usually requires further tests. Immunological techniques, DNA probes and PCR-based methods are mainly used for recognizing the pathogenic and non-pathogenic bacteria present in the coconut [12]. Molecular biological methods using PCR amplification of the bacterial 16S ribosomal ribonucleic acid (rRNA) gene using universal bacterial primers and sequencing are commonly used for accurate identification and classification of bacterial isolates [13]. This research was aimed at isolation, biochemical and molecular characterization of microorganisms in coconut which may have significant health concern.

2. METHODS AND MATERIALS

2.1 Materials and chemicals

The materials used in this research were: safety cabinet, laminar air flow, inoculation loops (spread and streak), erlenmeyer flask, bunsen burner, measuring cylinder, petri dish, test tubes, parafilm, plastic tape, autoclave, electric heater, electric balance, PCR thermocycler, PCR tubes, micropipettes, eppendrof tubes, electrophoresis equipments, GelDoc. The media and reagents used were: nutrient agar, Macconkey agar, Goat blood agar, distilled water, double distilled water, Genejet PCR kits, Geneaidg DNA isolation kit, TAE buffer (50X), agarose, Ethidium bromide (5mg/ml) [EtBr], ultrapure 6X grade gel loading dye.

2.2 Media preparation

As much as 52 gram of Macconkey agar was suspended in 1L of distilled water in erlenmeyer flask and mixed thoroughly; thenheatedfor 5 minutes to completely dissolve the powder. It was then autoclaved at 120°C for 15 minutes, 1 atm [14].

28 gram of nutrient agar was suspended in 1 litter of distilled water in another erlenmeyer flask and mixed by boiling to dissolve completely. It was then sterilized in an autoclave at 120°C for 15 minutes, 1 atm.

In the third erlenmeyer flask, 28g/l of nutrient agar was prepared, sterilized and goat blood agar was added to it when it cools to 40-50°C. All the Medias were allowed to cool at 40-50°C and transferred into different sterile petri-dishes for bacteria inoculation. The surfaces of the gels were dried for 2 days before inoculation.

2.3 Sample collection and bacteria Inoculation

Five samples from unfermented and three from fermented coconut were collected. The Unfermented were labeled as HR1, HR2, HR3, HR4 and HR5 whereas the fermented were labeled as with BP1, BP2, MC. All the petridish were labeled and spread technique was applied. Bacteria isolates were taken this previously enriched media using another loop and quadrant streaking technique was applied to isolate pure

colonies on petridish containing nutrient agar, Macconkey agar and Goat blood agar. The petri dishes were incubated for 24-48 hours at 37°C.

2.4 BIOCHEMICAL CHARACTERIZATION

A number of Biochemical tests were performed for the identification of bacterial isolates with the help of Oxoid manual [14]. Bacterial isolates were identified by their reaction to glucose, lactose, maltose, manitol, sucrose, Sulfide Indole Motility (SIM), oxidase and Simmon's citrate. To differentiate among *Enterobacteriaceae* group malonate broth, uria agar, Motility Indole Ornithine (MIO) and Lysine Iron agar (LIA) tests were conducted.

2.5 MOLECULAR CHARACTERIZATION

Molecular biological methods comprise a broad range of techniques that are based on the analysis and differentiation of microbial DNA. These methods test DNA directly from the microbial cells themselves and the nucleic acid can be cloned polymerase chain reaction (PCR), sequenced and analyzed.

2.5.1 DNA extraction

Bacterial genomic DNA was isolated based on Geneaidg DNA isolation kit. The detailed protocol is presented below in continuous steps:

2.5.1.1 Sample preparation

Up to 1 x 10^9 (1.5 ml) of bacterial cells were transferred from culture broth to a microcentrifuge tube. It was centrifuged at 14000xg for 1 minute and then the supernatant discarded. 200 µl of GT Buffer was added, vortexed for 5 seconds and incubated at room temperature for 5 minutes.

2.5.1.2 Lysis

200 µl of GB Buffer was added to the sample and mixed by shaking for 5 seconds. It was incubated at 60°C for 10 minutes to ensure the sample lysate is clear. During incubation, the tube was inverted and shaken every 3 minutes for mixing. At this time, the required Elution Buffer (200 µl per sample) was pre-heated to 60°C (for next step, DNA Elution).

2.5.1.3 DNA binding

200 µl of absolute ethanol was added to the sample lysate and mixed immediately by

votexing for 5 seconds. GD Column was placed in a 2 ml Collection Tube. All the mixture was transferred into the GD Column and then centrifuged at 14000xg for 2 minutes. The 2 ml Collection tube containing the supernatant was discarded. Then the GD Column was placed in a new 2 ml Collection Tube.

2.5.1.4 Washing

400 μl of W1 Buffer was added to the GD Column then centrifuged at 14000 x g for 30 seconds. The supernatant was discarded. The GD Column was placed back in the 2 ml Collection Tube. 600 μl of Wash Buffer two was added to the GD Column and centrifuged at 14000 x g for 30 seconds. Then supernatant was discarded. The GD Column was placed back in the 2 ml Collection Tube. It was centrifuged again for 3 minutes at 14000 x g to dry the column matrix.

2.5.1.5 Elution

The dried GD Column was transferred to a clean 1.5 ml microcentrifuge tube. 100 μl of the previously heated elution buffer was added. It was incubated at room temperature for at least 3 minutes to allow the Elution Buffer to be completely absorbed. It was centrifuged at 14000 x g for 30 seconds to elute the purified DNA. The DNA extraction result was kept at -21°C for the next procedures used in the next days.

2.5.2 Biophotometer DNA (Measuring DNA quantity and quality)

50 μl of double distilled water (ddH$_2$O) was added into cuvette as a blank control. The Cuvette was then placed in the cuvette shaft of the biophotometer. The blank button was pressed and checked if the result is 0.000A. 5 μl of ddH$_2$O was pipetted from the cuvette and discarded. Then 5 μl of DNA sample from test tube with BP$_2$ label was added to the cuvette containing ddH$_2$O and mixed by micro-pipetting. The cuvette was placed in the cuvette shaft of the biophotometer. The sample button was pressed to display the result and the result was recorded. All the above procedures were repeated for the other samples.

2.5.3 16S rRNA PCR amplification

PCR technology is the simplest and currently the most widely used method to obtain 16S rRNA genes for detailed characterization of microbial communities by amplifying a single target molecule by millions-fold. PCR methods are highly sensitive and specific. It is the foundation upon which most genetic polymorphism-based techniques are based [13].

The key to PCR is the use of oligonucleotide primers designed to be complementary to the desired gene or genetic region. Polymerase Chain Reaction (PCR) amplification was done using a pair of universal primers (63f:5'-CAGGCCTAACACAT-GCAAGTC-3' and 1387r: 5'-GGGCGGAWGTGTACAAGGC-3') for amplification of 16S rRNA [15].

The extracted DNA products were used as DNA template for amplifying 16S rRNA gene. 25 μl PCR mix include: 1.25μl each (forward and reverse) primers, 2 μl DNA sample, 12.5 μl Kapa (which includes Tag polymerase, PCR buffer, MgCl2, dNTPs premixed together) and 8 μl of double distilled water (ddH$_2$O). The conditions of PCR were as follows: initial denaturation for 3 minutes at 95°C, denaturation for 15 sec at 95^{0}C, annealing for 15 sec at 58°C, extension for 5 sec at 72°C, final Extension for 5 minutes at 72°C, Post PCR at 4°C, 35 cycles, cover temperature of 105°C.

2.5.4 DNA electrophoresis and GelDoc

The PCR products were separated by 1% w/v gel agarose electrophoresis, a technique used to separate biological molecules such as DNA, RNA, Proteins, Enzymes, etc. based on their migration ability, size, charge and pH. Generally, horizontal electrophoresis is used for nucleic acids (DNA, RNA) while vertical electrophoresis is applied for proteins including Isozymes [16].

Tris-acetate-EDTA (TAE) buffer was provided as stock of 50X, therefore it should be diluted due to the reason that failure in diluting the buffer will result in slow migration of samples and causes excessive heating of gel. Dilution was done to get 1L of 1X TAE buffer based on the following formula:

V_1C_1 $=V_2C_2$
V_1x50 $=1000$mlx1
V_1 $=20$ ml, hence 20 ml of 50X TAE was diluted in 980 ml distilled water.

1% w/v gel agarose was prepared by suspending 1 gram of agarose in 100ml of TAE (1X) buffer and dissolved by heating to boiling. It was then allowed to cool at 40-50^{0}C for almost 20 minutes. The two dams were placed in to the slots on each side of the gel plate and made tight. The melted agarose was poured onto the gel plate electrophoresis box. Gel comb with eight teeth (well maker) was inserted in a gel. The comb was nearest to the cathode (negative) as DNA migrates towards anode (positive). About 250 ml of TAE (1X) buffer was poured.

5 μl of 1Kb DNA marker plus 2 μl loading dye was used as ladder in the first well. 3 μl of PCR products were loaded in each of the other wells. DNA electrophoresis was run for 50 minutes at 90V, 400mA using thermo-scientific electrophoresis machine. The electrophoresis result was visualized by staining with ethidium bromide for 15-20 minutes followed by washing (to detoxicate ethidium bromide) with distilled water for 10-15 minutes and illuminated with UV light Gel Doc.

2.5.5 DNA sequencing, BLAST analysis, Multiple sequences Alignment

Four carbohydrate non-fermenting bacterial 16S rRNA isolates were sent to 1st BASE laboratories, Singapore, for 16SrRNA sequencing. 16S rRNA molecule has major role of protein synthesis in microorganisms. The mechanism of protein synthesis doesn't vary significantly from one organism to another due to the reason that the nucleotide sequences of some portions of 16S ribosomal deoxyribonucleic acid (rDNA) are sealed which are responsible for conserved nature of 16S rRNA. Nucleotide sequencing of this molecule is used to identify organisms and rank biological phylogenetic nomenclature by comparing certain locations on the 16S rRNA molecule with database of previously identified micro-organisms whose 16S rRNA mark is already known [12].

The 16S rRNA sequences were analyzed using Basic Local Alignment Search Tool (BLAST: http://blast.st-va.ncbi.nlm.nih.gov/Blast.cgi) from National Center for Biotechnology Information (NCBI) to find local similarity between sequences. This program compares sequence database and calculates the statistical significance of matches. Multiple sequence alignment using CLUSTALW2 program from EMBL European Bioinformatics Institute (EMBL-EBI: http://www.ebi.ac.uk/Tools/msa/clustalw2/) was performed to obtain phylogenetic tree that would help species relationship and evolutionary trends.

3. RESULT AND DISCUSSION

3.1 Biochemical Test Result

Test tubes containing carbohydrate broths (glucose, lactose, maltose, manitol and sucrose) were inoculated with bacteria and incubated at 37°C for 24 h. After incubation, a positive result was noted as change of color from red to yellow (MC, BP1 and BP2) whereas no color was observed in negative results (HR1, HR2, HR3, HR4 and HR4) as shown in Fig. 1 and Table 1.

Phenole red broths of crabohydrates are used to determine the ability of organisms to ferement carbohydates. The change in color from red to yellow indicates an acidic pH change, which is a positive indicator for carbohydrae fermentation. For glucose fermenting bacteria gaseous byproducts are formed and a bubble was detected in inverted tubes.

For the citrate utilizing test, Simons citrate agar slants were inoculated and incubated at 37°C for 24 h. All bacteria isolates were positive as color change from green to blue was noted due to drop in pH. For urease test urea broth was used and change in color from yellow-orange to bright pink was considered as positive. A positive oxidase by *Pseudomonas*, *Ralstonia pickettii and Burkholderia* indicated as development of dark purple color. This test was used to assess bacteria isolates which produce the enzyme cytochrome c-oxidase, a component of electron transport chain of some specific organisms.

Sulfide-Indole-Motility (SIM) medium allows detecting three different traits (sulfide production, indole production from tryptophan and bacterial motility). The hydrogen sulfide (H_2S) production allows differentiation of family *Enterobacteriaceae* as positive organisms reduce sulfur into H_2S and black precipitate was formed due to the reaction of hydrogen sulfide with ferrous sulfate. Indole test is used to detect organism's ability to breakdown Tryptophan using tryptophanase into indole and pyruvic acid. Indole binds with Kovac's reagent p-dimethylaminobenzaldehyde in isoamyl alcohol. Positive Indole test result forms a pink red layer (complex) on the top of the broth due to the reaction between indole and the reagent.

Motility Indole Ornithine Medium (MIO Medium) was used as differential test medium to detect the ability of bacteria to produce indole from tryptophan, to decarboxylate the ornithine and exibit motility by turbidity throughout of medium which enables the identification of *Enterobacteriaceae* on the basis of motility, indole production and ornithine decarboxylase activity. Motile organisms show either diffused growth or turbidity extending away from the inoculation line; while nonmotile organisms grow along the inoculation line. Organisms ferment dextrose to form acid which cause the bromo cresol purple which is the pH indicator to change from purple to yellow. Organisms which possess ornithine decarboxylase are able to decarboxylate ornithine to putrescine which increases the pH making it alkaline. The presence of casein enzymatic hydrolysate produces indole.

Lysin Iron Agar (LIA) test was used to differentiate gram negative bacteria to decarboxylate or deaminate lysine on the basis of lysine decarboxylase/deaminase and produce hydrogen sulfide (detected by ferric Ammonium Citrate indicator). L-Lysine helps to detect these two enzymes and sodium Thiosulfate acts as source of inorganic sulfur.

Based on Complete Biochemical test (tests used to further differentiate member of *Enterobacteriaceae* which include malonate, Uria, MIO and LIA), Sample BAP2 was detected as *Enterobacter agglomerans*, with results showing positive for carbohydrate fermentation, Citrate, Lysine Iron agar tests and H_2S production (black precipitate); but negative for malonate broth, urease and Ornithine tests.

On the other hand, MC1 and BP1 samples were detected as *Klebsiella Pneumonia*, with positive results for carbohydrate fermentation, citrate, malonate broth, uria agar, but negative for Sulfide Indole Motility (SIM) test, Motility Indole Ornithine (MIO) and H_2S production.

Samples HR1, HR2, and HR4were identified as *Ralstonia pickettii*, whereas sample HR3was identified as *Burkholderia* sp., (obtained from 16S rRNA sequence BLAST analysis) and Sample HR5 had negative results for carbohydrate and SIM test and but positive for citrate test; and was identified as *pseudomonas*.

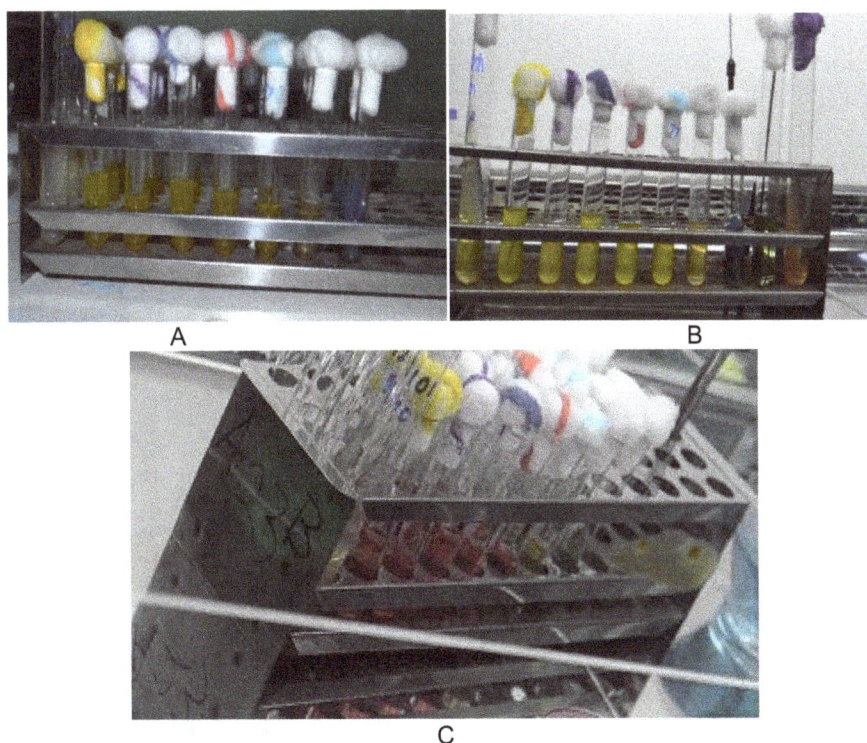

Fig. 1. A) Biochemical test I (for genera identification of bacteria isolates form fermented coconut) B) Biochemical test II for species identification of bacteria isolates form fermented coconut C) Biochemical test I (for genera identification of bacteria isolates form unfermented coconut

Table 1. Biochemical test results of bacteria isolates from fermented and unfermented bacteria

Tests		Samples sources								Remark
		Fermented coconut			Unfermented coconut					
		MC1	BP1	BP2	HR1	HR2	HR3	HR4	HR5	
Glucose		+g	+g	+g	-	-	-	-	-	+=yellow
Lactose		+	+	+	-	-	-	-	-	g=gas
Maltose		+	+	+	-	-	-	-	-	formation
Manitol		+	+	+	-	-	-	-	-	
Sucrose		+	+	+	-	-	-	-	-	
SIM	Sulfide	-	-	+	-	-	-	-	-	
	Indole	-	-	+	-	-	-	-	-	
	Motility	-	-	+	-	-	-	-	-	
Simmon's citrate		+	+	+	+	+	+	+	+	+=blue
Oxidase test		NA	+	+	+	+	+	+	+	
Malonate broth		+		-						
Uria agar (urease test)		+		-	NA					
MIO	Motility	-		+						
	Indole	-		+						
	Ornithine	-		-						
LIA (Lysine Iron agar)	Slant	K	K	K						
	Butt	A	A	A						
	H₂S	-	-	+						

NA= Not applied

All these species are related to human pathogenicity. This can be related to hygiene in food processes and handling procedures. Therefore, to favour the selective growth of beneficial bacteria, appropriate environmental hygiene is required.

3.2 Molecular Characterization Result

3.2.1 DNA extraction and Biophotometer analysis

When checked by Eppendrof biophotometer the value of the bacterial DNA isolates at purity level of A260/280 ratio ranged from 1.88 to 2.0 (Table 2). DNA isolates are said to be pure and eligible to proceed to the molecular analysis if the A260/280 ratio values ranged from 1.8 to 2.0.

3.2.2 16S rRNA PCR amplification, DNA electrophoresis and GelDoc

PCR amplification of bacterial 16SrRNA gene using bacterial universal oligo-primers followed by sequencing of the PCR amplicons which help to determine bacterial species available in coconut. Depending on the data of nucleic acids obtained from PCR amplification of the bacterial 16S rRNA, the bacterial variants would be investigated. [12]. This produces a variety of community profiles in which complex banding patterns of DNA fragments on a gel, commonly referred to as a "fingerprint", represent the diversity of polymorphic genes of interest, which in turn reflects the diversity of microbial species in the community. [13].PCR amplified product sizes were determined by comparing the length of the DNA bands migrating with DNA markers of known size and concentration after running DNA electrophoresis. The results of 16S rRNA gene amplification with primers 63F and 1387r are found to be approximately 1,300 bp each as seen in a DNA marker size 1 Kb as shown in Fig. 2.

The denaturation of double stranded DNA at high temperature of 95°C separates the DNA into single stranded nucleotides. The two oligonucleotide primers (forward and reverse) anneal complementary regions of the denatured DNA at 58°C. Finally a heat stable DNA polymerase created new stands of DNA for each template by using the complementary strand as a template to extend the primer.

Table 2. Yield of genomic DNA from 1 x 10^9 cells from fermented an unfermented coconut extracted by genomic DNA extraction kit (bacteria)

Sample source	Bacterial isolates	DNA concentration (µg/ml)	The ratio A260/A280
Fermented coconut	BP2 (1st replica)	3.1	1.91
	BP2 (2nd replica)	3.6	1.88
	MC1 (1st replica)	6.5	1.95
	MC1 (2nd replica)	3.8	1.97
Unfermented coconut	HR1	7.3	1.99
	HR2	3.1	1.93
	HR3	8.2	1.96
	HR4	5.0	2.0
	HR5	7.6	1.98

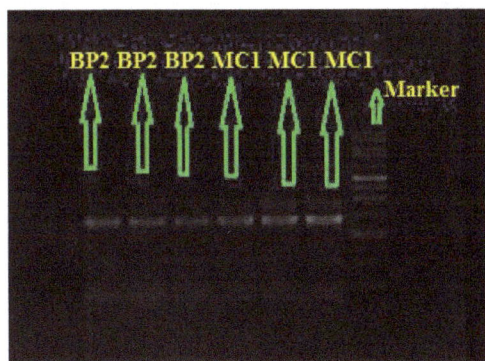

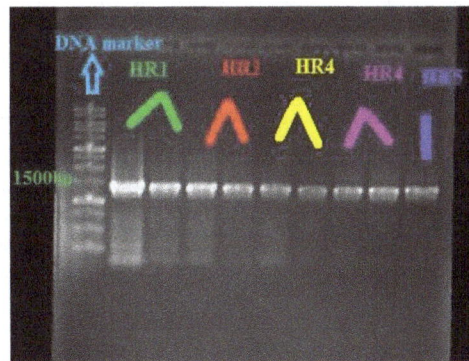

Fig. 2. DNA electrophoresis analysis of PCR products; the 16S rRNA products (amplified by using 63f and 1387r primers) were obtained from genomic bacteria DNA isolated from both fermented and unfermented coconut. The PCR products were separated by 1% w/v gel agarose electrophoresis

3.2.3 16S rDNA sequencing, BLAST analysis and Multiple Sequence Alignment

The sequences of four carbohydrate non-fermenting bacteria isolates were obtained from first BASE laboratories, Singapore. By using universal PCR primers complementary to conserved regions of the rRNA gene and subsequent cloning and sequencing of the PCR products, the 16S rRNA gene sequences are compared with other known 16S rRNA sequences to identify the bacteria species. The identity of some of the bacteria were determined by sequencing and BLAST analysis as shown in Table 3. 16S rRNA sequences enable the identification microorganisms because the fixed 16S rRNA contains species-specific variable regions that change according to different species [12].

Ralstonia pickettii is an infrequent invasive pathogen, but can cause infections, mainly of the respiratory tract, in immunocompromizd and cystic fibrosis patients [17]. *Ralstonia pickettii* and Ralstonia insidiosa were reported [18] as emerging waterborne bacteria pathogens and have survival capability in various water sources and grow in various water sources. They were found to be susceptible to antibiotics such as quinolones and sulfamethoxazole/trimethoprim. Apart from pathogenicity, *Ralstonia pickettii* possess a significant Biotechnological potential for bioremediation due to its ability to break down many toxic substance and xenobiotic pollutants such as toluene and trichloroethylene [19]. Some species of *Burkholderia* such as *Burkholderia pseudomallei* and *Burkholderia mallei* were associated to melioidosis and glanders [20,21].

The comparison of 16SrRNA sequence enables to construct phylogenetic trees which make it easier for studies of species diversity and molecular evolution as shown in Fig. 3. For recognition of closest species, the obtained sequences were match up with 16S rRNA gene sequence in GenBank databases using the National Center for Biotechnology Information [12]. The 16S rRNA was thus used for phylogenetic analysis as well as for the species-specific detection of bacteria. The partial 16S rRNA gene sequences are used in constructing phylogenetic trees based on CLUSTALW program to determine identity or nearest phylogenetic position.

Table 3. Identification of bacteria based on BLAST analysis of 16S rRNA sequence data

Isolate code	16S rDNA size (bp)	Similarity from BLAST	Accession	identity
HR1	1262	*Ralstonia pickettii* strain S-X17A 16S ribosomal RNA gene, partial sequence	KJ806364.1	97%
HR2	*1255*	*Ralstonia pickettii* strain VIT-SRM1 16S ribosomal RNA gene, partial sequence	KJ716446.1	98%
HR3	*1270*	*Burkholderia sp.* B38 16S ribosomal RNA gene, partial sequence	KF788057.1	97%
HR4	*1255*	*Ralstonia pickettii* strain S-X17A 16S ribosomal RNA gene, partial sequence	KJ806364.1	96%

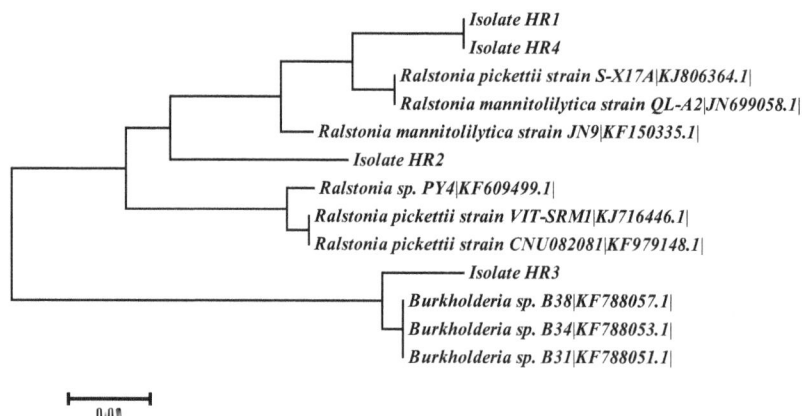

Fig. 3. Phylogenetic tree of 16S rRNA nucleotide sequence of bacteria from coconut; phylogenetic tree is based on 1300 bp (63f-1387r)

Based on phylogenetic tree analysis Isolate HR1 is 97% similar with | Ralstoniapickettii strain S-X17A, 97% similar with Isolate HR2, 96% similar with Isolate HR4 and 91% similar with Isolate HR3. Isolate HR2 is 98% similar with Ralstoniapickettii strain VIT-SRM1, 95% similar with Isolate HR4, and 91% similar with Isolate HR3. Isolate HR3 is 97% similar withBurkholderia sp. B38 but 90% similar with Isolate HR4. Isolate HR4 is 96% similar with Ralstoniapickettii strain S-X17A.

4. CONCLUSION

The combined application of biochemical and molecular techniques ensured to identify the bacteria species found in both fermented and unfermented coconut. Accordingly, *Klebsiella pneumonia* and *Enterobacter aggromelans* were found in fermented coconut whereas *Pseudomonas*, *Ralstonia pickettii* and *Burkholderia* were detected in unfermented coconut. BLAST followed by Multiple Sequence Alignment enabled to construct phylogenetic tree which helped to study the similarity and evolutionary relationship among the bacteria isolated from coconut.

COMPETING INTERESTS

Authors have declared that no competing interests exist.

REFERENCES

1. Khan MN, Rehman MU, Khan KW. A study of chemical composition of *Cocos nucifera* L (Coconut) water and its usefulness as rehydration fluid, Pak. J. Bot. 2003;35(5):925-930.
2. Prades A, Dornier M, Diop N, Pain JP. Coconut water uses, composition and properties: a review, Journal of Fruits. 2012;67:87–107.
3. Bourdei R, Konan JL, N'Cho YP. Coconut: a guide to traditional and improved varieties, Ed. Diversiflora, Montpellier, France; 2005.
4. Ediriweera ERHSS. Medicinal uses of coconut (*Cocos nucifera* L.), Coco info Int. 2003;10:11–21.
5. Krishna AG, Raj G, Bhatnagar AS, Prasanth Kumar PK, Chandrashekar P. Coconut Oil: chemistry, production and its applications - A Review, Indian Coconut Journal. 2010;15-27.
6. Igbabul BD, Bello FA, Ani EC. Effect of fermentation on the proximate composition and functional properties of defatted coconut (*Cocos nucifera* L.) flour. Sky Journal of Food Science. 2014;3(5):034–040.
7. Tripetchkul S. Utilization of wastewater originated from naturally fermented virgin coconut oil manufacturing process for bioextract production: Physico-chemical and microbial evolution. Journal of Bioresources Technology. 2010;101:6345-6353.
8. Kabara JJ. Fatty acid and derivates as antimicrobial agents A review, in the Pharmacological effect of lipids (Kabara JJ, ed.). American Oil Chemist Society, Campaign IL; 1995.
9. Kumalaningsih S, Padaga M. The utilization of microorganisms isolated from fermented coconut milk for the production of virgin coconut oil, J. Basic. Appl. Sci. Res. 2012;2(3):2286-2290.
10. Isaacs CE, Thomas H. The role of milk derived antimicrobial lipids as antiviral and antibacterial agents. Adv Exp. Med. Bio. 1991;310:159-165.
11. Meroth BC, Hammes WP, Hertel C. identification and population dynamics of yeasts in sourdough fermentation processes by pcr-denaturing gradient gel electrophoresis. Appl. Environ. Microbiol. 2003;69(12):7453-7461.
12. Geevarghese A, Baskaradoss JK, Al Dosari AF. Application of 16s rrna in identifying oral microflora-a review of literature. World Applied Sciences Journal. 2012;17(10):1303-1307.
13. Woo PC, Lau SK, Teng JL, Tse H, Yuen KY. Then and now: use of 16S rDNA gene sequencing for bacterial identification and discovery of novel bacteria in clinical microbiology laboratories: Review. Clin Microbiol Infect. 2008;14:908–934. DOI:10.1111/j.1469-691.2008.02070.x.
14. Bridson EY. The OXOID Manual. 8th Edition. OXOID Limited, Wade Road, Basingstoke, Hampshire, RG24 8PW, England. 1998;371.
    Available: http://www.oxoid.co.uk.
15. Marchesi JR, Sato T, Weightman AJ, Martin TA, Fry JC, Hiom SJ, Wade WG. Genes coding for bacterial 16s rrna bacterium-specific pcr primers that amplify design and evaluation of useful. Appl. Environ. Microbiol. 1998;64(2):795.
16. Hailu Weldekiros Hailu, Danang Hari Kristiyanto, Ali Ramadan Ali ALatawi,

Saleh Muhammed Raqib. Isozyme electrophoresis and morphometric comparison of reed (*Imperata cylindrical*) adaptation to different altitudes. International Journal of Innovative Research in Science, Engineering and Technology. 2014;3(5):12387-12394.

17. Stelzmueller I, Biebl M, Wiesmayr S, Eller M, Hoeller E, Fille M, Weiss G, Lass-Floerl C, Bonatti H. Ralstoniapickettii—innocent bystander or a potential threat? Clin Microbiol Infect. 2006;12:99–101. DOI: 10.1111/j.1469-0691.2005.01309.x.

18. Ryan MP, Adley CC. The antibiotic susceptibility of water-based bacteria *Ralstonia pickettii* and *Ralstonia insidiosa*. Journal of Medical Microbiology. 2013. 62:1025–1031.
DOI 10.1099/jmm.0.054759-0.

19. Ryan MP, Pembroke JT, Adley CC. *Ralstonia pickettii* in environmental biotechnology: potential and applications: Review Article. Journal of Applied Microbiology. 2007;103:754–764. DOI:10.1111/j.1365-2672.2007.03361.x.

20. Currie BJ, Fisher DA, Anstey NM, Jacups SP. Melioidosis: acute and chronic disease, relapse and re-activation. Trans. R. Soc. Trop. Med. Hyg. 2000;94:301-304.

21. Hsueh PR, Teng LJ, Lee LN, Yu CJ, Yang PC, Ho SW, Luh KT. Melioidosis: an emerging infection in Taiwan? Emerg. Infect. Dis. 2001;7:428433.

8

Determination of Starch and Cyanide Contents of Different Species of Fresh Cassava Tuber in Abia State, Nigeria

O. R. Ezeigbo[1*], C. F. Ukpabi[2], C. A. Ike-Amadi[2] and M. U. Ekaiko[1]

[1]Deparment of Biology/Microbiology, Abia State Polytechnic, Aba, Nigeria.
[2]Deparment of Chemistry/Biochemistry, Abia State Polytechnic, Aba, Nigeria.

Authors' contributions

This work was carried out in collaboration between all authors. Author ORE designed the study, wrote the protocol and managed the analyses of the study. Authors CFU and CAI wrote the first draft of the manuscript and managed the literature searches while author MUE performed the statistical analysis. All authors read and approved the final manuscript.

<u>Editor(s):</u>
(1) Yan Juan, Doctorate of Horticultural Crop Breeding, Sichuan Agricultural University, Ya'an, China.
<u>Reviewers:</u>
(1) Kolawole O.P, Ibadan, Nigeria.
(2) Sanmei MA, Department of biotechnology, Jinan University, The People's Republic of China.

ABSTRACT

This investigation aimed at determining the cyanide and starch contents of different species of cassava grown in Abia State. Samples of cassava tubers were purchased from the local market and identified by a taxonomist from the National Root Crops Research Institute, Umudike, Abia State. In this study, the cyanide levels obtained were less than 100 mg/kg for both sweet and bitter cassava. The lower concentration of cyanide obtained from this survey, could have been influenced by the season of harvest (rainy season) in Nigeria. Of the six species analyzed, "0581" recorded the highest in cyanide content with 62.57 ± 0.10 mg/kg, followed by "30211" with 59.55 ± 0.19 mg/kg. Sweet cassava (0505) recorded the least with 36.65 ± 0.16 mg/kg cyanide. On the starch content, 21.70 ± 0.10% was obtained from "30572", followed by "8083" with 20.62 ± 0.11% and "30211" had the least starch content with 17.48 ± 0.02%. Both Starch (%) and HCN (mg/kg) are statistically significant (*p-value* < 0.001) in the means for the various species of cassava under study. However, from the posthoc test, there is no significant difference between species 30211 and 0581. The other species are significantly different at 5% level. However, all the species of

---

Corresponding author: E-mail: obyezeigbotxt1@yahoo.com

cassava tested had cyanide above the recommended level (10 mg/kg). Consumption of these cassava species unprocessed/inadequately processed would lead to serious health challenges and therefore, efforts are required to reduce cyanide content at least to the recommended level.

Keywords: Cassava tubers; Starch; Cyanide; Abia State.

1. INTRODUCTION

Various poisonous substances that affect human metabolism have been identified in some edible plants. One of such poisonous substances is linamarin, a cyanogenic substance that is highly concentrated in cassava [1]. Cassava (*Manihot escilanta, crantz*) constitutes one of the major stable foods for an estimated 500 million people in the tropical world and 700 million people worldwide [2,3,4,5]. The high carbohydrate content makes cassava a major food item especially for the low income earners in most tropical countries especially Africa and Asia [6]. The main food source is the starchy roots while the leaves are processed and used as sources of protein in Africa [7]. The tubers are processed into gari, fufu, chips and flour for baking bread and biscuits [4,8,9]. Thus, cassava serves as a major source of dietary energy but deficient in protein [10].

Nigeria ranked first among major producers of cassava in the world, producing about 33 million metric tones of cassava in 1993 [11,12]. Over 95% of cassava produced in Nigeria is used as food for the country's teeming population, being one of the countries staple foods [11]. However, the presence of cyanogenic glycosides (linamarin) in cassava has greatly affected utilization of cassava as food source. The cyanogenic glycosides form effective biological system to deter predators [13]. Cyanide binds to the enzyme, cytochrome oxidase which negatively affects respiration. Akintonwa et al. [14] had earlier reported cases of consuming cassava diets containing high levels of dietary cyanide. Non-fatal doses of cyanide have been associated with such symptoms like dizziness, headache, stomach pains, vomiting and diarrhea which appear four or more hours after consuming poorly processed cassava [15]. In some cases, excessive consumption of cassava due to improper preparation has been linked to such ailments like goiter, ataxia (a neurological disorder affecting the ability to walk) [16,17], chronic pancreatitis [18] and even death.

Toxic levels of cyanogenic glycosides have been used as the basis for categorizing cassava varieties. The less toxic varieties are designated as sweet cassava (*Manihot palmate*) while the more toxic varieties are designated as bitter cassava (*Manihot utilissima*). The bitter variety is widely cultivated because of its high yield and economy Onwuka A. Production and characterization of adhesive from cassava. B. Eng. Research Project. Department of Chemical Engineering, Delta state University, 2012, Unpublished], has high concentration of cyanogenic glycoside (100-500 mg/kg fresh weight), which is distributed throughout the tuber. The sweet variety has low concentration of cyanogenic glycoside (less than 100 mg/kg) which is confined mostly to the peels [19]. Cyanide level of less than 50 mg/kg is slightly poisonous, 80-100 mg/kg is toxic and levels above 100 mg/kg are fatal especially in grated cassava [20]. In most cases, the cyanogens in fresh cassava roots are detoxified during processing into edible products [5,21]. The processing of cassava roots commences with the peeling of the tubers, which contains higher cyanide content than the pulp. Removal of the peels therefore reduces the cyanogenic glycoside content considerably [22]. Thus, the selection of cassava species to be grown is very important. This study aimed at determining the cyanide and starch content of varieties of cassava grown in Abia State, Nigeria. This will assist farmers in their choice of cassava for cultivation.

2. MATERIALS AND METHODS

2.1 Sample Collection and Preparation

Fresh cassava tubers were obtained from the local market and identified by the taxonomist from the National Root Crops Research Institute, Umudike, Abia State and transported to the National Research Laboratory for analysis. Some cassava tubers were cleaned and peeled using a knife and later washed with clean water before analysis.

2.2 Determination of Hydrogen Cyanide

The hydrogen cyanide content of the fresh samples of cassava tuber was determined using the method described by Onwuka [23]. 5 g of each cassava variety was blended into a paste and dissolved in 50 mL of distilled water in a conical flask and allowed to stay overnight before filtering. 2 mL of the filtrate was poured into another conical flask and 4 mL alkaline picrate solution was added and the content incubated in a water bath for 5 minutes for colour change (reddish-brown) and absorbance taken at 450 nm. The blank and standard test (which serves as control) was prepared using 1 mL distilled water and 4 mL alkaline picrate solution. The cyanide content was extrapolated using a cyanide standard curve.

Calculation: 100/W × An/As × C × Vf/Va × 1/106 (mg/kg)

Where Vf= total volume of extract
Va= volume of extract used
An= Absorbance of standard
As= Absorbance of standard
C= concentration of standard

2.3 Determination of Starch Content

The percentage content of starch was determined using the spectrophotometer method as described by Radley [24]. 2.5 g of ground fresh cassava tubers was dissolved in 50 mL of cold water and allowed to stand for an hour. 20 mL of HCl and 150 mL of distilled water were added to the sample and refluxed for 2 hours in a round bottom flask. This was later cooled and neutralized with 0.5 N NaOH. The content was used for glucose determination using anthrone reagent.

Series of glucose solution was prepared containing 0 ppm, 2 ppm, to 10 ppm to calibrate the glucose standard concentration. 5 mL anthrone reagent was added to each of the standard and test sample and boiled in a water bath for 20 minutes for colour development. The test tubes were cooled and absorbance was taken at 620 nm against a black containing only 1 mL of water and 5mL of anthrone reagent.

Calculation: Absorbance/volume ×SR ×50/25 ×100

Where SR = slope reciprocal
Volume= volume of sample used.

2.4 Statistical Analysis

The data was analyzed using Statistical Package for Social Sciences (SPSS) version 20.0. The statistical tools employed are descriptive statistics, Analysis of Variance (ANOVA) and Duncan Multiple range test. The descriptive statistics was used to determine the various measures of central tendency. The means of the various cassava species were compared using ANOVA, and the Duncan Multiple range test was used to make a pair-wise comparison of all the possible pairs of means. Statistical significance tests included the use of *p-value* to assess for the role of chance. In this study, p-value = 0.05 was used to disapprove the null hypothesis.

3. RESULTS

Six species of cassava were analyzed. 30572 species recorded the highest starch value of 21.70%, followed by 8083 species with 20.62% while 30211 species had the least starch of 17.48% (Table 1). The cyanide content shows that 0581 species had the highest with 62.57 mg/kg, followed by 30211 with 59.55 mg/kg while sweet cassava (0505) had the least cyanide content of 36.65 mg/kg. Both Starch (%) and HCN (mg/kg) are statistically significant (*p-value* < 0.001) in the means for the various species of Cassava under study. However, from the posthoc test, there is no significant difference between species 30211 and 0581. The other species are significantly different at 5% level. Figs. 1 and 2, representing the mean plot of starch and cyanide respectively, provide quick appreciation of the data.

4. DISCUSSION

The results yielded concentrations of cyanide ranging from 36.65 mg/kg to 62.57 mg/kg of fresh tuber. Except the sweet cassava (0505), all other species could be considered rich in cyanide. The analysis revealed higher levels of cyanide above the recommended safe limit of 10 mg/kg [25]. Bitter cassava (30211, 30572, 0581 and 8083) species recorded higher concentration of cyanide than the sweet cassava (0505). Wheatley et al. [19] obtained similar results, although the present survey recorded lower concentrations of cyanide in all the species investigated. According to literature, time of harvest influences cyanide content of fresh cassava. Harvest conducted during the rainy season and in the afternoon would significantly reduce the rate of cyanide in the cassava products [26]. In rainy season, the cyanogenic glycosides, which are soluble in water, are partially released into the soil [27]. This property is used in the technique of soaking for the elimination of cassava cyanide [28, 29].

Table 1. Starch and cyanide content of different species of cassava

Species of cassava	Starch (%) –Mean ± SD	Cyanide (mg/kg)- Mean ± SD
0505 (SC)	18.59. ± 0.17	36.65 ± 0.16
30211 (BC)	17.48 ± 0.02	59.55 ± 0.19
30572 (BC)	21.70 ± 0.10	49.86 ± 0.57
0581 (BC)	17.47 ± 0.23	62.57 ± 010
8083 (BC)	20.62 ± 0.11	52.68 ± 0.10

Key: 0505 = sweet cassava; 30211, 30572, 0581 and 8083 = Bitter cassava; SC= sweet cassava; BC= bitter cassava

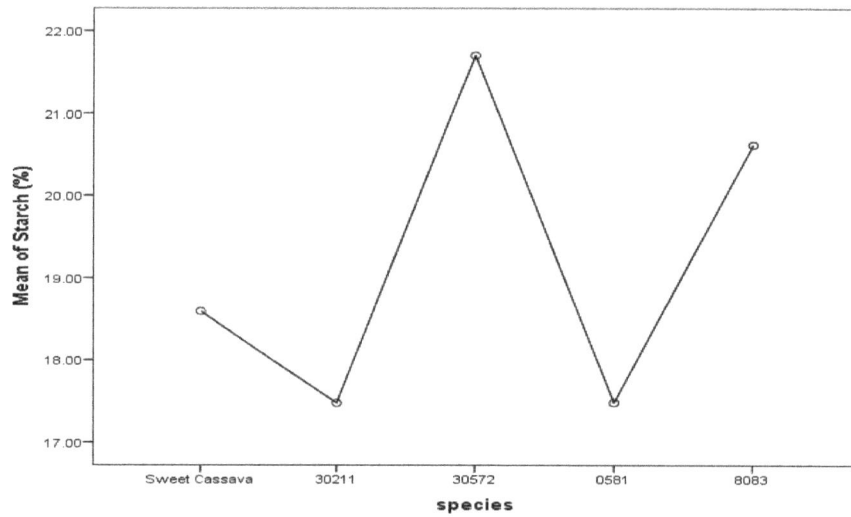

Fig. 1. Mean plot for starch

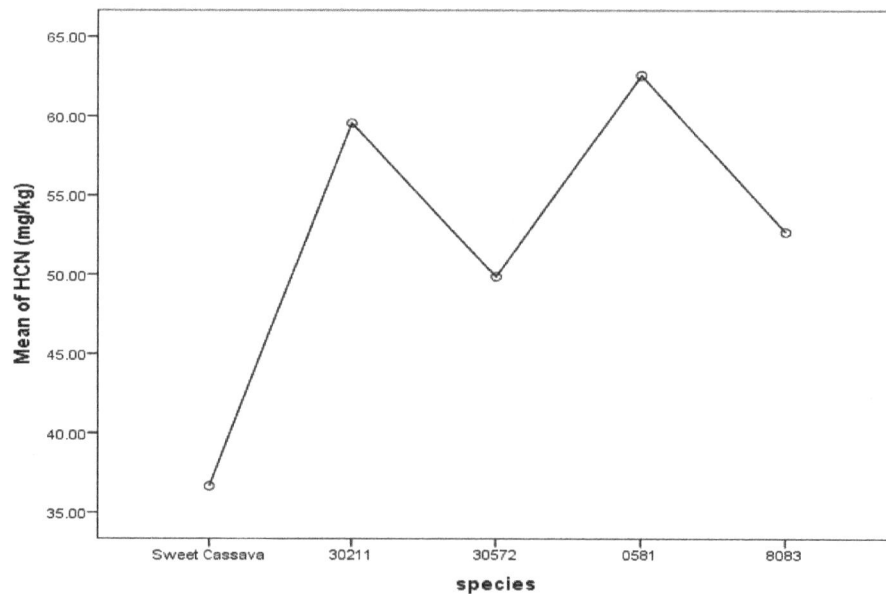

Fig. 2. Mean plot for cyanide

Cyanide level of less than 50 mg/kg, is slightly poisonous, 80-100 mg/kg is toxic and level above 100 mg/kg is fatal especially in grated cassava [20]. From this study, only sweet cassava falls under the category of being slightly poisonous; the others (bitter cassava species) are categorized as toxic with levels above 50 mg/kg [30]. However; the bitter cassava is cultivated more as a result of its high yield and economy [Onwuka A. Production and characterization of adhesive from cassava. B. Eng. Research Project. Department of Chemical Engineering, Delta state University, 2012, Unpublished]. This is confirmed in this study as two species of bitter cassava studied (30572 and 8083) recorded higher starch contents of 21.70% and 20.62% respectively when compared to the sweet cassava (18.59%). Although, the cyanide levels obtained in this study fall below the toxic levels of 80-100 mg/kg, epidemiological studies have shown that exposure of small doses of cyanide over a long period can result in increased blood cyanide levels [31,32].

5. CONCLUSION

The presence of cyanogenic glycoside has greatly affected the consumption of cassava. In this study, the cyanide levels obtained were less than 100mg/kg for both sweet and bitter cassava. The lower concentration of cyanide obtained from this survey, could have been as a result of the washing down of the cyanide into the soil by rain water, since the cassava was harvested during the rainy season. However, efforts are still required to reduce the cyanide level further below the harmful limits.

ACKNOWLEDGEMENT

The team is grateful to the National Root Crops Research Institute, Umudike, Abia State in identifying the cassava species.

COMPETING INTEREST

Authors have declared that no competing interests exist.

REFERENCES

1. Nwabueze TU, Odunsi FO. Optimization of process conditions from cassava (*Manihot esculenta*) lafan producntion. African Journal of Biotechnology. 2007;6(5):603-611.
2. International Institute of Tropical Agriculture (IITA). Cassava in tropical Africa: A reference manual. International Institute of Tropical Agriculture, Ibadan. 1990;87-120.
3. FAO. The State of Food Security in the World. Food and Agricultural Organization, Rome, Italy; 2003.
4. Oboh G, Akindahunsi AA. Biochemical changes in cassava products (flour and garri) subjected to saacharomyces cerevisiae solid media fermentation. Food Chemistry. 2003;82:599-602.
5. Udoudoh PJ. Post harvest storage and spoilage of cassava tuber (*Manihot* spp) in Ikot Ekpene, Akwa Ibom State, Nigeria. Journal of Environmental Issues and Agriculture in developing counties. 2011;3(3):34-35.
6. Desse G, Taye M. Microbial loads and microflora of cassava (*Manihot esculenta* crantz) and effects of cassava juice or some food borne pathogens. J. Food Test Afri. 2001;6(1):21-24.
7. Bradbury JH. Simple wetting method to reduce cyanogens content of cassava flour. Journal of Food Composition and Analysis. 2006;19(4):388-393.
8. Oti EE. Acute toxicity of cassava effluent to the African catfish fingerlings. Journal of Aquat. Sc. 2002;17:31-34.
9. Cardoso AP, Mirione E, Ernesto M, Massaza F, Cliff J, Haque M, Bradbury JH. Processing of cassava to remove cyanogens. Journal of Food Composition and Analysis. 2005;18:451-460.
10. FAO Food and Agriculture Organization, Production Year Book. 1984;38:130. FAO Rome.
11. Onwuka GI, Ogbogu NJ. Effects of fermentation on the quality and physicochemical properties of cassava based fufu products made from two varieties, NR8212 and Nwangbisi. Journal of Food Technology. 2007;5(3):261-264.
12. Akinpelu AO, Amangbo LEF, Olojede AO, Oyekale AS. Health implications of cassava production and consumption. Journal of Agriculture and Socialb Research. 2011;11(1):118-120.
13. Mkpong OE, Yan H, Chism G, Sayre RT. Purification, characterization and localization of linamarase in cassava. Plant Physiology. 1990;93:176-181.
14. Akintonwa A, Tunswaashe O, Onifade AA. Fatal and non-fatal acute poisoning attributed to cassava based meal. Acta Horticul. 1994;375:285-288.
15. Osuntokun BO. Ataxic neuropathy associated with high cassava diets in West Africa. In chronic cassava toxicity. Proc. Interdisciplinary workshop (Nestel B and MacIntyre R. eds) London, IDRC Ottawa; 1973. IDRC-010e.

16. Food and Agriculture Organization of the United Nations (FAO). Root tubers, plantains and bananas in human nutrition, Rome, Toxic substances and anti-nutritional factors. 1990;7. Available: http://www.fao.org/docrep/t0207e/T0207E0 8.htm#Cassava%20toxicity. (Accessed 18 August 2014.).

17. Bhatia E. Tropical calcific pancreatitis: strong association with SPINK1 trypsin inhibitor mutation. Gastroenterology. 2002;123:1020-1025.

18. Wagner H. Cassava's cyanide-producing abilities can cause neuropathy; 2005. Available:http://www.cidpusa.org/cassava. htm. Retrieved 23 July, 2014.

19. Wheatley CC, Orrego JI, Sanchez T, Granados E. Quality evaluation of cassava core collection at CIAT. In WM Rica and AM Thro (eds), Proceedings of the First International Scientific Meeting of the Cassava Biotechnology, New York, CCIAT, Crtagena, Columbia. 1993;255-257.

20. Okaka JC, Okaka ANC. Food composition, spoilage, shelf-life extension. OCJANCO Acad. Publishers, Enugu, Nigeria. 2001;225-226.

21. Okaka JC. Tropical plant perishables, handling, storage and processing. Silicon Valley Publishers, Enugu. 1997;122.

22. Onwuka GI. Food Analysis and Instrumentation (Theory and Practice). 1st Edn., Napthali Prints, Surulere, Lagos-Nigeria. 2005;140-160.

23. Radley JA. Starch production technology. London: Applied Science Publishers Ltd. 1976;203-208.

24. FAO/WHO Codex Alimentarius Commission. Report of the eight session of the Coordinating Committee for Africa, Cairo, Egypt; 1988.

25. Tchacondo T, Karou SD, Osseyi E, Agban A, Bawa LM, Atcha AR, Soumana K, Assignon K, Kpemoua A, De Souza C. Effect of harvest and cook processing on ncyanide content of cassava-based dough consumed in Lome, Togo. Advanced Journal of Food Science and Technology. 2011;3(5):399-403.

26. Silvestre P, Arraudeau M. Le manioc. In Techniques Agricoles et Production Tropicales. Maisonneuve et Larose, Paris, France. 1983 ;32:262.

27. Bradbury JH, Denton IC. Rapid wetting method to reduce cyanogens content of cassava flour. Food Chem. 2010;121:591-594.

28. Mlingi NL, Nkya SR, Tatala S, Rashid S, Bradbury JH. Recurrence of Konzo in southern Tanzania: Rehabilitation and prevention using the wetting method. Food Chem. Toxicol. 2011;49:673-677.

29. James CS. Analytical Chemistry of Foods. Champion and Hall, Glass. 1995:74.

30. Nambisan B, Strategies for elimination of cyanogens from cassava for reducing toxicity and improving food safety. Food Chem. Toxicol. 2011;49:690-693.

31. Soto-Blanco B, Maiorka PC, Gorniak SL. Effect of long term low- dose cyanide administration to rats. Ecotoxicology and Environmental Safety. 2002;53(1):37-41.

32. Rachinger J, Feller FA, Steiglbauer K, Trankler J. Epidemiological changes after acute cyanide in toxication. American Journal of Neuroradiol. 2002;23:1398-1401.

9

Compositional and Amino Acid Profile of Nicker Bean (*Entada gigas*) Seeds

H. N. Ogungbenle[1*] and O. T. Oyadipe[1]

[1]*Department of Chemistry, Ekiti State University, P.M.B. 5363, Ado - Ekiti, Nigeria.*

Authors' contributions

This work was carried out in collaboration between both authors. Author HNO designed the study, performed the statistical analysis, wrote the protocol, managed the literature searches and wrote the first draft of the manuscript. Author OTO managed the analyses of the study. Both authors read and approved the final manuscript.

Editor(s):
(1) Alok Adholeya, Biotechnology and Management of Bio-resources Division, The Energy and Resources Institute, India.
(2) Kuo-Kau Lee, Department of Aquaculture, National Taiwan Ocean University, Taiwan.
Reviewers:
(1) Fagbemi, Tayo Nathaniel, Food Science and Technology, Federal University of Technology, Nigeria.
(2) Anonymous, USA.
(3) Anonymous, Nigeria.
(4) Anonymous, Mexico.
(5) Anonymous, Brazil.

ABSTRACT

The proximate, anti-nutritional factors, functional properties, minerals and amino acid composition of nicker bean (*Entada gigas*) were determined. The sample contained crude protein and carbohydrate of 24.8±0.02% and 47.2±0.10% respectively. The sample exhibited good emulsion capacity and emulsion stability of 62.3±0.03% and 36.2±0.01% respectively. The foaming capacity was 35.0±0.02% while the least gelation concentration was 4.00±0.01%. It contained nutritionally valuable minerals. The values of the anti-nutritional factors were: phytate (6.59±0.50 mg/g), oxalate (3.50±0.10 mg/g) and tannin (0.26±0.02%). Glutamic acid was the most abundant amino acid with the value of 51.1 g/100 g crude protein while methionine (4.10 g/100 g crude protein) was the least abundant amino acid in the sample. *Entada gigas* is a potential source of nutrients that should be cultivated and its consumption encouraged.

Keywords: Proximate; anti-nutritional; functional; minerals; amino acid.

*Corresponding author: E-mail: httphenryo@yahoo.com

1. INTRODUCTION

Global alliance for improved nutrition is paramount to food scientists. Food scientists are always in search for newly improved and underutilized legumes in order to proffer solutions to the global food shortages and consequent malnutrition which has lead to series of prevalent diseases. *Entada gigas* has not been consumed as food in Africa but as medicinal plant for local treatment of some ailments. It is known as fever nut; spreading vine-like in nature which grows as tall as 6-7 m in length and uses other vegetation for support. The stem is covered by sharp spines which may grow to 0.4 or more metres in diameter [1] and the seed grows in a flat pod with length of 0.04 m. Two or three hard smooth seeds of 0.01 m in length are contained in each pod. The seeds inside pod are olive green in colour and later become dark-red brown at maturity. *Entada gigas* is very common in coastal habitats including beaches and dry areas in West Africa [1]. The data obtained from the analyses would help us to know whether the sample has good nutritional values as other existing edible plants or may be processed for future consumption because the seed is known in Africa to contain certain toxins which hinder its edibility. Functional properties of any food must be known before it can be incorporated into food system such as emulsification, foamability and gelation [2]. Therefore, the aim of this work is to determine the proximate, functional properties, anti-nutrients and amino acid composition of the sample.

2. MATERIALS AND METHODS

The sample used for the present work was bought in Ado-Ekiti central market, Ekiti State Nigeria in Africa continent. The 5 kg of the sample was separated from the shells, later dried and ground into flour. The flour was packaged and stored in freezer (-4°C) prior to the analyses.

2.1 Proximate Analysis

The moisture and total ash contents were determined using the air oven and muffle furnace as described by Pearson [3]. The sample was analyzed for crude fat and crude protein using methods of AOAC [4] and the percentage nitrogen was converted to crude protein by multiplying by 6.25. The crude fiber was determined by adding 2 g of the sample into

200 ml of 1.25% H_2SO_4 in 500 ml conical flask and boiled for 30 minutes. The mixture was later filtered through muslin cloth, rinsed with hot distilled water and scrapped back into the flask and added 200 ml of 1.25% NaOH and allowed to boil again for another 30 minutes. The mixture was filtered, rinsed with 10% HCl twice with industrial methylated spirit and allowed to drain and dry. The residue was scrapped into a crucible, later dried in an oven at 105°C, cooled in a desicator, re-weighed and finally incinerated in muffle furnace at 300°C for 30 minutes, again cooled at room temperature and re-weighed [4]. The carbohydrate value was calculated by difference.

2.2 Determination of Functional Properties

The water and oil absorption capacities of the sample were determined using the methods of Beuchat [5]. Ten milliliters was added to 1.0 g sample in a centrifuge tube. The suspension was mixed vigorously using Marlex mixer and then centrifuged at 15,000 rpm for 15 minutes and the volume of the supernatant was recorded. The bound water was calculated from the difference in the initial volume of the solvent used and the final volume after centrifuging. The same procedure was used for oil absorption capacity by replacing King's vegetable oil of density 0.880 g/ml with water in the method described above.

The emulsion capacity and stability were determined according to methods of described by Lin et al. [6]. Two grams sample flour was added to 100 ml distilled water and blended for 30 seconds using Marlex food mixer at a high speed. After complete dispersion, 5 ml portion of King's vegetable oil of 0.880 g/ml density was added from a burette with continuous blending until the emulsion break point (i.e. a separation into two layers) at room temperature. The value obtained was expressed as gram of oil emulsified per 1 gram of the sample.

The emulsion stability was determined as the volume of the water separated after 24 hours at room temperature.

The slightly modified method of Sathe et al. [7] was used to determine the least gelation concentration. Sample slurries of 2, 4, 6, 8, 10, 12, 14, 16, 18 and 20% were prepared in 5 ml portions of distilled water. The test tubes containing these slurries were heated for one hour in boiling water followed by rapid cooling for 2 hours at -4°C. The least gelation concentration was determined as concentration which did not slip when the test tubes were inverted.

The method of Coffman and Garcia [8] was employed to determine foaming capacity and stability. One gram of the sample was whipped with 50 ml distilled water for 5 minutes in a Marlex food mixer and later poured into a 100 ml graduated flask to study the foaming capacity (per cent increase in volume). The foaming stability was determined as the difference between the foam and the water level after 2 hrs.

2.3 Mineral Analysis

The minerals were analyzed by dry ashing the sample at 550°C in a Carbolite, Sheffield muffle furnace to constant weight and dissolving the ash in 100 ml standard flask using distilled deionized water with 3 ml of 3 M HCl. Sodium and potassium were determined using a flame photometer (model 405, corning, U.K). All other minerals were determined using Atomic Absorption Spectrophotometer (Perkin & Elmer model 403, USA) [9].

2.4 Determination of Anti Nutrients

Tannin: A 10 ml of 70% aqueous acetone was added to 200 mg of the sample, tightly covered and put into an ice bath then shaken for 2 hours at 30°C, later centrifuged at 3,600 rpm. A 0.2 ml of the mixture was pipetted into a test tube and 0.8 ml of distilled water was added. Standard tannic acid solutions were prepared from 0.5 mg/ml stock solution and made up to 1 ml with distilled water. Folin Wu reagent (0.5 ml) and 2.5 ml of 20% Na_2CO_3 were added to the sample mixture and the standard solutions; both solutions were incubated for 40 minutes at room temperature and the absorbance was measured at 725 nm [10].

Oxalate: One gram of the sample was weighed into 100 ml conical flask, 75 ml of 1.5 M H_2SO_4 was added and stirred for 1 hour then filtered. The sample filtrate (25 ml) was titrated against 0.1 $MKMnO_4$ solution until faint color persisted for 30 seconds [10].

Phytate was determined on Spectronic 20 colorimeter (Gallenkamp, UK). The amount of phytate in the sample was determined as hexaphosphate equivalent [11,12].

2.5 Determination of Amino Acid

The amino acid profile was determined using the method described [13]. The sample was dried to constant weight and defatted using Soxhlet extractor. After the defatting process, the defatted sample (2 g) was weighed into a glass ampoule; 7 ml of 6MHCl was added and oxygen was expelled by passing nitrogen into the ampoule so as to avoid possible oxidation of some amino acids during hydrolysis. The glass ampoule was then sealed with bunsen burner flame and placed in an oven at 105±5°C for 22 hours. The ampoule was allowed to cool before broken at the tip and the content was filtered to remove the organic matters. The filtrate was then evaporated to dryness at 40°C under vacuum in a rotavapor (BÜCHI Rotavapor, R110). The residue was dissolved in 5 ml of acetate buffer (pH 2.0) and stored in specimen bottles which were kept in the freezer. The hydrolysate (7.5 µL) was dispensed into the cartridge of the Technicon Sequential Multi-Analyser (TSM) using a syringe. The TSM analyser is designed to separate and analyse neutral, acidic and basic amino acids of the hydrolysate. The amount of amino acid was obtained from the chromatogram peaks. The whole analysis lasted for 76 minutes and the gas flow rate was 0.50 ml/min at 60°C with reproducibility consistent within ±3%.

3. RESULTS AND DISCUSSION

3.1 Proximate Composition

Proximate composition of the *Entada gigas* is presented in Table 1. The crude fat content of the *Entada gigas* was lower than those of legume seeds [14] and *Colocynthis citrullus* (51.9%) [11]. The value of carbohydrate was 47.1±0.10% and that of total ash was 2.55±0.12%. The crude protein was higher than that of cereal, millet (11.4%) [15]. But this present value was higher than those of lima bean flour (22. 7%) [16], pigeon pea (22.4%) [16], African mango full fat (10.6%) and defatted (14.8%) [17]. The sample contained crude protein of 24.8±0.02%. The high crude protein content may be valuable for food formulation. The moisture content (8.94±0.02%) was higher than those of *Parinari curatillefolia* (4.40%) [18], full fat (5.76%) and defatted (5.44%) of cashew nut kernel [19]. The crude fiber content (4.87±0.01%) was lower than that of *Parinari curatillefolia* (5.45%) [18], while the crude fat content of the sample (15.7±0.03%) was higher than those of *L. leucocephalia* (6.80%) [20] and scarlet runner bean (5.3-6.9%) [21].

Table 1. Proximate composition of
Entada gigas

Parameter	%
Moisture	8.94±0.02
Total ash	2.55±0.12
Crude fat	15.7±0.03
Crude fiber	4.87±0.01
Crude protein	24.8±0.02
Carbohydrate by difference	47.2±0.10

3.2 Functional Properties

The results of functional properties are presented in Table 2. The emulsion capacity for *Entada gigas* is 62.3±0.03% while the emulsion stability was 36.2±0.01%. The value of the emulsion capacity was higher than those of benniseed (30%), cowpea (28.0%) [22], dehulled African nutmeg (45.6%) [23], dehulled African yam bean (10.0-20.0%) [24] and soy flour (18.0%) [6]. It implies that *Entada gigas* may be good for food additives and emulsified foods like ice cream and salad dressings. The foaming capacity (35.0±0.02%) was lower than those of *Luffa cylindrica* (60%) [25] and *African nutmeg* (50%) [23]. The foaming stability was lower than that reported for *African nut meg* (8.5%) [23]. The foaming capacity and stability make the sample good for food formulations and stabilizing colloidal food system.

Table 2. Functional properties of
Entada gigas

Parameter	%
Water absorption capacity	145±0.20
Oil absorption capacity	250±0. 30
Foaming capacity	35.0±0.02
Foaming stability	5.00±0.15
Emulsion capacity	36.2±0.01
Emulsion stability	62.3±0.03
Least gelation concentration (W/V)	4.00±0.01

The least gelation concentration was 4.00±0.01%. The ability of the bean to form gels, provides a structural matrix for holding water, flavor, sugar and food ingredients is useful in food application and in new product development, thereby providing an added dimension to food functionally [26]. The least gelation concentration observed may be an asset

in the use of the *Entada gigas* for formulation and as an additive to other gel- foaming material in food products [27]. This value was lower than those of pigeon pea (12%) [2], benniseed [15], lupin seed flour (14%) [28], *Afzelia africana* (6%) [29] but higher than that of African nutmeg [23]. This indicates that the sample may be a useful additive to other materials for gel formation in food products.

The oil absorption capacity (250±0.30%) was higher than those of benniseed (45.5%) [15], millet 55% [30] and conophor nut (108.13%) [14], and lower than that of African nutmeg (256%) [23].

3.3 Mineral Composition

The results of the mineral analysis of *Entada gigas* are presented in Table 3. The highest concentrated mineral element was sodium with the value of 362±0.30 mg/100 g followed by calcium (361±0.20 mg/100 g) while selenium (1.00±0.03 mg/100 g) was the least abundant. This indicates that the value of calcium in the sample is adequate for infant development of bones and teeth. Therefore, suggests that *Entada gigas* flour might be a good supplement for human food formulations in future if accepted as food.

The value of calcium (361±0.20 mg/100 g) in *Entada gigas* was higher than those of dehulled African nutmeg (203.7 mg/100 g) [31] and *Afzelia africana* (208 mg/100 g) [32]. The value of Fe (10.5±0.04 mg/100 g) was also higher than that of dehulled African nutmeg (3.0 mg/100 g) [31]. Fe is an important component in the transfer of oxygen and it is most closely related to anaemia. Iron is highly useful for the blood formation [22]. The Na^+/K^+ ratio is an indicator for mineral recommendation. The Na/K ratio of 1.0 is recommended as reported by Niemann [33]. Na and K control water equilibrium level *in* body tissue and are also important in the transportation of some non-electrolytes. Mg also activates enzymatic process in the body. Cu and Zn are known to be associated with enzyme systems particularly the biochemical oxidation-reduction processes. The results showed that the sample is rich in most of the essential minerals and could therefore serves as additional source for such minerals. On comparative basis, *Entada gigas* is a better source of Ca, Mg, Zn. Mg and Fe respectively.

Table 3. Mineral composition of *Entada gigas*

Mineral	mg/100 g
Fe	10.5±0.04
Zn	75.8±0.01
Cu	1.33±0.02
Mn	25.7±0.10
Ca	361±0.20
Na	362±0.30
K	62.2±0.08
As	N.D
Cd	N.D
Ni	1.20±0.10
Se	1.00±0.03
Pb	1.50±0.01
Co	N.D

N.D – Not Detected

3.4 Anti-nutritional Factors

The results of oxalate, tannin and phytate contents are presented in Table 4. The phytate level (6.59±0.50 mg/g) was higher than those of sorghum (5.34 mg/g) and millet (4.41 mg/g), Afzelia *africana* (13.59%) [29] and date palm fruit (4.87 mg/g) reported by Ogungbenle [34]. Oxalate (3.50±0.10 mg/g) *was* lower than those of sorghum (5.22 mg/g) and millet (4.06 mg/g) [35] but higher than those of *Afzelia africana* (2.84 mg/g) [29] and date palm fruit (3.0 mg/g) [34]. The value of tannin (0.26±0.02%) was lower than those of date palm fruit (3.0%) [34], millet (4.06%) [12], sorghum (0.44%) [35] and cooked walnut (2.33%) [36]. Tannin has been shown to contribute some degrees of resistance to pre-harvest germination [37]. The high value of phytate, indicates that the *Entada gigas* may be processed chemically as a source of phytic acid, which binds polyvalent cations and makes them unavailable in the body. Tannin which was very low makes the sample important medicinally as a detersive and astringent in intestinal tubules [38]. It can also prevent alcohol intoxication in the body system.

3.5 Amino Acid Profile

The result of the amino acid analysis of the *Entada gigas* is presented in Table 5. Glutamic acid (Glu) was the highest amino acid with the value of 51.1 g/100 g while methionine was found to be least concentrated amino acid with the value of 4.10 g/100 g. This is in agreement with the observation of Olaofe et al. [39]. The glutamic acid value of the sample was higher than those of cooked walnut (151.6 g/100 g) [36], raw millet (67.4 g/100 g) [40], white melon (180.9

g/100 g) and bulma cotton flour (189.9 g/100 g) [41]. The total essential amino acids (TEAA) of the sample amounted to 206.3 mg/g (with histidine), 199.3 mg/g (without histidine) (Table 5). The TEAA (with histidine) value of the *Entada gigas* was lower than that of soy bean; 444 g/100 g [42]. *Entada gigas* is also a good source of leucine (Leu) and arginine (Arg). Table 5 also shows that *Entada gigas* has high quality protein and a good source of essential amino acid for children [42,43]. One can also conclude that the sample will also meet the amino acid requirement for adult since the amino acid composition that meets the requirement for a child will also meet that for an adult [43,44]. Maize food products are highly deficient in lysine and they are the most common weaning foods for children in most African countries [45], *Entada gigas* may not be therefore suitable for the enrichment of maize food products since it contains low amount of lysine (12 g/100 g). The quality of dietary protein can be measured in various ways [46] but basically as the ratio of the available amino acids in the food or diet compared with the body needs, expressed as a ratio [47-49]. Methionine and cysteine were the limiting amino acids in the present sample. This was also observed by Salem [50] for faba bean materials. Apata and Ologhobo [51] also reported low amounts of methionine and cysteine in legumes. The values of these amino acids were higher than those of faba bean materials [50] and velvet tamarind pulp methionine (4.30 g/100 g), cysteine (5.83 g/100 g) [52].

Table 4. Anti-nutritional factors of
Entada gigas

Parameter	Value
Tannin (%)	0.26±0.02
Phytate (mg/g)	6.59±0.50
Oxalate (mg/g)	3.50±0.10

The percent TNEAA was 35.6% and percent TEAA was 64.4% (with histidine), 62.2% (without histidine) showing that TNEAA was less concentrated in the sample than TEAA (%TNEAA< %TEAA), thus making it a good source of essential amino acids (Table 6). The % TEAA for the sample (64.4%, with histidine) and (62.2%, without histidine) was higher than those of germinated millet (50.2%, with histidine) and (46.3%, without histidine) [40] and *Afzelia africana* (48.5%, with histidine) and (41.6%, without histidine) [29]. Histidine is an important amino acid required by children for growth and development.

Table 5. Amino acid composition of
***Entada gigas* (g/ 100 g crude protein)**

Amino acid	g/100 g crude protein
Lysine*	12.0
Histidine*	7.00
Arginine*	46.1
Aspartic acid*	23.0
Threonine*	6.60
Serine	12.0
Glutamic*	51.1
Proline	6.10
Glycine	15.6
Alanine	10.0
Cystine	5.50
Valine	23.0
Methionine*	4.10
Isoleucine*	11.2
Leucine*	50.2
Tyrosine	13.6
Phenylalanine*	23.1

• *Essential amino acid*

**Table 6. Essential, non essential, neutral,
acidic and basic amino acids**

TAA	320
TNEAA	114
TEAA (with histidine)	206
TEAA (without histinine)	199
% TNEAA	35.6
% TEAA (with histidine)	64.4
% TEAA (without histidine)	62.2
TNAA	62.2
% TNAA	58.7
TAAA	74.1
% TAAA	23.1
TBAA	58.1
% TBAA	18.1

TAA   -   *Total Amino Acid;*
TNEAA - *Total Non Essential Amino Acid*
TEAA   -   *Total Essential Amino Acid*
TNAA   -   *Total Non Amino Acid*
TAAA   -   *Total Acidic Amino Acid*
TBAA   -   *Total Basic Amino Acid*

4. CONCLUSION

This work has revealed that the *Entada gigas* has nutritive values which are compared favorably with other sources of protein. It can also be deduced from the results that the sample has high quality essential minerals and high amount of anti-nutrient factors. Hence, its consumption is encouraged in future provided the anti-nutrients are reduced through processing.

COMPETING INTERESTS

Authors have declared that no competing interests exist.

REFERENCES

1. Irvine R. West African Crops, 3rd edition. London Oxford University, Press; 1980.
2. Oshodi AA, Ekperigin MM. Functional properties of pigeon pea (*Cajanus Cajan*) flour. Food Chem.1989;34:187-191.
3. Pearson D. Chemical analysis of foods. 7th ed. Churchill Livingstone. London; 1979.
4. AOAC. Official methods of analysis 15th edition. Association of Official Analytical Chemists Washington D.C.; 2005.
5. Beuchat LR. Functional and electrophoretic characteristics of succylated peanut flour. J Agric Food Chem. 1977;25:258 -261.
6. Lin MYY, Humbert ES, Sosulki FO. Certain functional properties of sunflower meal products. J Food Sci. 1974;39:368-370.
7. Sathe SK, Desphande SS, Salunkhe DK. Functional properties of winged bean protein. J Food Sci. 1982;47:503-509.
8. Coffman C, Garcia VV. Functional properties of protein isolate from mung bean flour. J Food Tech. 1977;12:437-484.
9. Ogungbenle HN. Chemical and fatty acid composition of *Afzelia africana* seeds, Adv Analyt Chem. 2014;4(2):30-34. DOI:10.5923/j.aac20140402.02.
10. Day RA (Jnr.), Underwood AL. Quantitative analysis. 5th ed. Prentice Hall Publication. 1986;701.
11. Ogungbenle HN, Oshodi AA, Oladimeji MO. Chemical energy evaluation of some underutilized legume flours. Riv Italia Sos Grasse. 2005;82(4):204-208.
12. Harland BF, Oberleas D. Phytate content of foods: Effect on dietary zinc bioavailability. J Am Diet Assoc. 1981;79:433-436.
13. Spackman DH, Stein EH, Moore S. Automatic recording apparatus for use in the chromatography of amino acids. Analyt Chem. 1958;30:1191.
14. Ige MN. Functional properties of the protein of some Nigeria oil seeds. Conophor seeds and three varieties of some Nigeria oil seeds. Food Chem. 1984;32:822-825.
15. Oshodi AA, Ogungbenle HN, Oladimeji MO. Chemical composition, functional properties and nutritionally valuable

minerals of benniseed, pearl millet, quinoa seed flours. Int J Food Sci Nutri. 1999;50:325-331.

16. Oshodi AA, Adeladun MOA. Proximate composition, some valuable minerals and functional properties of three varieties of lima bean flour. Int J Food Sci Nutri. 1993;43:181-185.

17. Ogungbenle HN. Chemical and amino acid composition of raw and defatted African mango kernel. Brit Biotechnol J. 2014;4(3):244-253.

18. Ogungbenle HN, Atere AA. The chemical, fatty acid and sensory evaluation of *Parinari curatellifolia* seeds. Brit Biotechnol J. 2014;4(4):379-386.

19. Ogungbenle HN. Chemical functional properties and amino acid composition of raw and defatted cashew kernel. Am Chem Sci J. 2014;4(3):348-356.

20. Adewuyi A, Oderinde RA. Characterization and fatty acid distribution in the lipids of *Leucaena leucocephala* and *Prosopis africana*. Pak J Sci Ind Res. 2012;55:86-91.

21. Aremu MO, Olaofe O, Akintayo ET. Nutritional quality assessment of the presence of hull in some Nigeria underutilized legume seeds. Bull Pure Appl Sci. 2005;24:47-52.

22. Ogungbenle HN. Nutritional evaluation of quinoa flour. Int J Food Sci Nutri. 2003;54:153-158.

23. Ogungbenle HN, Adu T. Proximate composition and functional properties of dehulled African nutmeg. Pak J Sci Ind Res. 2012;55:80-85.

24. Oshodi AA, Ipinmoroti KO, Adeyeye EI. Functional properties of some varieties of African Yam Bean (*Sphenostylis stenocarpa*) flour. Int J Food Sci Nutri. 1997;48:243-250.

25. Olaofe O,Yinka O, Aremu MO. Chemical evaluation of the nutritive value of smooth *Luffa cylindrica* seed's kernel. Electr J Food Chem. 2008;7:3444-3452.

26. Oshodi AA, Olaofe O, Hall GM. Amino acid fatty acids and mineral composition of pigeon pea. Int J Food Sci Nutri. 1993;43:187-191.

27. Altschul AM, Wilcke AL. New protein foods; Food science and technology. Academic Press Orlando FL; 1985.

28. Sathe SK, Salunkhe DK. Functional properties of lupin seed and protein concentrate. J Food Sci. 1981;46:149-197.

29. Ogungbenle HN, Omaejalile M. Functional properties and anti nutritional properties, *in-vitro* protein digestibility and amino acid composition of dehulled *A. africana* seeds. Pak J Sci Ind Res. 2010;53:265-270.

30. Ajayi IA, Dawodu FA, Adebowale KO, Oderinde RA. A study of the oil content of Nigerian grown *Monodora myristica* seeds for its nutritional and industrial application. Pak J Sci Ind Res. 2004;47:60-65.

31. Ogungbenle HN. Chemical, *In vitro* digestibility and fatty acids composition of African nutmeg. Ann Sci Biotechnol. 2011;2:46-51

32. Ogungbenle HN. Chemical and fatty acid composition of *Afzelia africana*. Adv Analyt Chem. 2014;4(2):30-34. DOI:10,5923/j.20140402.02.

33. Niemann DC, Butterworth DE, Niemann CN. Nutrition-Winc. Brown Publication, Dubugue U.S.A. 1992;237-312.

34. Ogungbenle HN. Nutritional quality and amino acid composition of date palm fruit flour. Riv Italia Sos Grasse. 2008;85:54-59.

35. NAS. Recommended dietary allowances. National Academy of Science-National Research Council. 8th ed. Washington DC; 1974.

36. Ogungbenle HN. Chemical and amino acid composition of cooked walnut flour. Pak J Sci Ind Res. 2009;52:130-133.

37. Hulse JH, Laing EM, Pearson OE. Sorghum and the millets: Their Composition and Nutritive Values. Academic Press London; 1980.

38. Morton M. Tamarind: In fruits of warm climates. Florida flair books, Miami. Web; 1987. Available:http://www.hort.Purdue.edu/newcrop/morton/tamarind html

39. Olaofe O, Adeyemi OF, Adediran GO. Amino acid and mineral composition and functional properties of oil seeds. J Agric Food Chem. 1994;42:878-884.

40. Adeyeye EI. Intercorrelation of amino acid quality between raw steeped and germinated pearl millet grains. Pak J Sci Ind Res. 2009;52:122-129.

41. Ogungbenle HN. Chemical composition, functional properties and amino acid composition of some edible seeds. Riv Italia Sos Grasse. 2006;83:71-79.

42. Altschul AM. Proceedings of plant protein foodstuff. Academic Press New York USA. 1958;713-714.

43. WHO. Energy and protein requirement. WHO Technical Report Series No. 522. Geneva WHO; 1973.

44. FAO/WHO. Energy and protein requirement report of joint FOA/WHO expert consultation. FAO Foods and Nutrition Paper 51 FAO/WHO Rome Italy; 1985.

45. Akinrele IA, Edwards CA. An assessment of the nutritive value of a maize soy mixture as a weaning food in Nigeria. Brit J Nutri. 1971;26:177-185.

46. FAO/WHO. Protein quality evaluation, report of joint FAO/WHO expert consultation. FAO Food and Nutrition Paper 51. Rome Italy; 1970.

47. Orr ML, Watt BK. Amino acid content of foods. Economics Research Report 4, USDA. Washington D.C. USA; 1957.

48. FAO. Utility of tropical foods: Tropical beans. Food and Agriculture Organization Publication. 1970;22-26.

49. Bender A. Meat and meat products in human nutrition in developing countries. FAO Food and Nutrition Paper 53. Rome, Italy; 1992.

50. Sale SA. Chemical composition of faba bean genotypes under various water regimes. Pak J Nutri. 2009;8(4):477-482.

51. Apata DF, Ologhobo AD. Biochemical evaluation of some Nigerian legume seeds. Food Chem. 1994;49:333-338.

52. Ogungbenle HN, Ebadan P. Nutritional qualities and amino acid profile of velvet tamarind (Dialium guineense) Pulp. Brit Biomed Bull. 2014;2(1):6-16.

10

Optimization of Polyphenols Extraction Method from Kola Nuts (*Cola nitida* Vent. Schott & Endl.) Using Experimental Design

Yves B. Nyamien[1,2*], Olivier Chatigre[1], Emmanuel N. Koffi[3], Augustin A. Adima[2], and Henri G. Biego[1]

[1]Laboratory of Biochemistry and Food Science, Training and Research Unit of Biosciences, Felix HOUPHOUËT-BOIGNY University of Abidjan, 22 BP 582 Abidjan 22, Côte d'Ivoire.
[2]Laboratory of Water Chemistry and Natural Substance, Training and Research Department of GCAA, Felix HOUPHOUËT-BOIGNY National Polytechnic Institute, BP 1093 Yamoussoukro, Côte d'Ivoire.
[3]Laboratory of Bioorganic Chemistry and Natural Substance, Training and Research Unit of Applied Fundamental Science, Nangui ABROGOUA University of Abidjan, 02 BP 801 Abidjan 02, Côte d'Ivoire.

Authors' contributions

This work was carried out in collaboration between all authors. Author YBN designed the study, performed the statistical analysis, wrote the protocol, and wrote the first draft of the manuscript. Authors OC, ENK, AAA and HGB managed the analyses of the study and literature searches. All authors read and approved the final manuscript.

<u>Editor(s):</u>
(1) Anil Kumar, School of Biotechnology, Devi Ahilya University, Madhya Pradesh, India.
<u>Reviewers:</u>
(1) Nii Korley Kortei, Department of Nuclear Agriculture and Radiation Processing, Graduate School of Nuclear and Allied Sciences, University of Ghana, Ghana.
(2) Anonymous, Croatia.

ABSTRACT

Aims: The aim of this study is to optimize extraction process of phenolics compounds from kola nuts by using experimental design.
Study Design: Kola nuts were collected in October 2014-February 2015 in south of Côte d'Ivoire. Harvested kola nuts were transferred to the laboratory until used in the experiments
Place and Duration of Study: This study was carried out during season 2014-2015 in the Laboratory of Biochemistry and Food Science, Félix Houphouët-Boigny University, Côte d'Ivoire.

Corresponding author: Email: nyams02@gmail.com

Methodology: Nuts were divided into two groups and subdivided to obtain four groups according to their variety (traditional or improved) and morphotype or cultivar (white and red). After drying and powder processing, the effect of six parameters (solvent type, solid-liquid ratio, extraction mode, variety, cultivar and extraction time) on polyphenol extraction from kola nuts were studied. Firstly, a Plackett-Burman design (8 experiments) was used to highlight the most important factors which influence the extraction process. Then, a full factorial design (2^k, k=4) was used to optimize the extraction conditions.

Results: Results showed that solvent, ratio (w/v), extraction mode and variety of nuts had significant effect on polyphenols extraction. The predicted optimal conditions for the highest polyphenol content from kola nuts were found with infusion of traditional variety at 1/100 (w/v) ratio with aqueous ethanol 50% (v/v). In the predicted optimal conditions, experimental values were 350 mg/L GAE, 1460 mg/L QE and 264.33 mg/L CE for total polyphenol, total flavonoid and condensed tannin, respectively. Experimental data were very close to the predicted values.

Conclusion: The extractive capability of kola nuts polyphenol is considerably depended on the solvent type, the extraction mode, the solid-liquid ratio (w/v) and nuts variety. Thus, kola nuts can be considered as a natural source of phenolics compounds with good antioxidant capacity. This optimization of the extraction parameters of phenolics compounds from kola nuts is very original, it is the first on a current scale of research on kola nut.

Keywords: C. nitida; polyphenols; optimized extraction; full factorial design.

1. INTRODUCTION

Natural antioxidants are increasingly appreciated by consumers due to both their inherent positive effects and to the possibility of using them as a source of natural additives to replace synthetic ones [1,2]. They are radical scavengers which protect the human body against free radicals that may cause pathological conditions such as ischaemia, asthma, arthiitis, inflammation, neurodegeneration, etc. [3]. The search of new antioxidants and phenolics from natural herbal source has taken very large attention in last decade [4]. Antioxidant activity of fruits and vegetables is generally positively correlated with their content of polyphenols [5]. Polyphenols are widely distributed in plants with antioxidant and antiradical properties [6,3], although they are found in relatively high amount in some plants, seeds and fruits [7] and beverages [8]. They have been studied by reserchers for their strong antioxidant health benefits [9-11].

Natural and secondary metabolic substances, especially, polyphenols, represent a wide range of substances with various structures, fall into different families including anthocyanins, coumarins, lignins, flavonoids, tannins, quinones, acids and phenols [5,6]. This structural diversity results in large variability of the physico-chemical properties influencing polyphenol extraction. Experimental data *in vitro* suggest that polyphenols have anti-inflammatory, antiallergic, anti-viral, and anti-carcinogenic activities [12-14]. They also have a protective role against chronic degenerative diseases (cataracts, macular degeneration, neurogenerative diseases), cancer cardiovascular diseases [13].

Over the years, several assay methods, solvents and sample preparation techniques have been adopted for the quantification of polyphenolics but there is no universal extraction procedure suitable for extraction of all plant phenolics [15]. Extraction methods (traditional methods, ultrasound-assisted extraction, subcritical water extraction, supercritical fluid extraction, pressurized fluid extraction, or accelerated solvent extraction), particle size, sample preparation, extraction time, solvent type, temperature of extraction and the presence of interfering factors have been shown to strongly influence polyphenol extractability in plant materials [16-19].

Kola nuts is a natural plant that is a rich source of natural polyphenols and provides a high free radical scavenger activity [20-23]. There are over 140 species of kola nut trees, and the most commonly edible are bitter cola (*Garcina kola*), kola nuts (*C. nitida* or *C. acuminata*) [20,24,25]. These three species are used as stimulants, increasing energy and strenght, dispelling drowsiness and staving hunger [26,27,22]. But only *C. nitida* is one of great interest because locally cultivated, widely consumed when fresh, while the dried nuts are used for beverages and pharmaceutical purposes in Europe and North America [21,18].

Many parameters have significantly influenced the extraction yield of kola nuts. In order to extract the bioactive polyphenolic compounds from *Cola nitida* nuts, an experimental design with a Plackett-Burman design following by a full factorial design was carried out. We have used firstly the screening designs (Plackett-Burman deign) which are best known for factors two levels. The experimentation highlights in a large number, those factors that are actually influential on a process in a fixed experimental field [28-29]. The choice of screening design is based on the number of the studied factors which is six (extraction method, extraction solution, solid-liquid ratio (w/v), extraction time, color and variety of nuts). The most important factors acting on bioactive compounds extraction were used in a second experimental design, full factorial design, that to examine the interactions effects of the factors on a response or dependent variable by carry out all possible combinations of levels and variable [30].

According to the litterature, no work has so far been reported on the optimization of extraction of bioactive compounds from Côte d'Ivoire kola nuts. Under this situation a statistical method of optimization seems to be very useful. The objective of the present study was to optimize extraction conditions by using non-toxic solvent for the enhanced recovery of polyphenols from kola nuts. Plackett-Burman or Hadamard experimental design allowed the screening to identify the most important factors among ratio, extraction method, solvent type, extraction time, nut variety and morphotype or cultivar [29]. After this screening, the optimization was done by full factorial design while keeping in mind the factors that influence the extraction along with the effects of interaction between these factors [31].

2. MATERIALS AND METHODS

2.1 Plant Material

A fresh kola nut was used as plant material. They were collected from October 2014 to February 2015 in south of Côte d'Ivoire.

As shown in Table 1, harvested plant materials were organized into two groups and subdivided to obtain four groups according to their color or cultivar and their variety. Two varieties of nuts

are available in Côte d'Ivoire: nuts did not undergo genetic modification (Traditional or spontaneous nuts) and those from the breeding program initiated by the National Agricultural Research Centre (improve nuts). At laboratory, they were washed with distilled water, cut into smaller pieces and dried at room temperature ($30\pm2^{\circ}C$) during two weeks. The dried sample was milled into powder using an electric blender, and stored in plastics bags prior to analysis.

Table 1. Sampling of kola nuts

Sample identification	Fresh nuts color	Variety
RCN_1	Red	Traditional
RCN_2	Red	Improve
WCN_1	White	Traditional
WCN_2	White	Improve

RCN : Red C. nitida, WCN : White C. nitida

2.2 Chemical Reagent

All reagents used in the study were of pure analytical grade, unless otherwise specified. Ethanol (CH_3CH_2OH), methanol (CH_3OH), hydrochloric acid (HCl), potassium peroxodisulfate ($K_2S_2O_8$), sodium nitrite (Na_2NO_2), Folin-Ciocalteu's phenol reagent, sodium carbonate salt (Na_2CO_3), sodium hydroxide (NaOH) and aluminum chloride ($AlCl_3$) were purchased from Carlo Erba (Spain). Trolox (6-hydroxy-2,5,7,8-tetramethylchroman-2-carboxylic acid), gallic acid, quercetin, catechin were purchased from Sigma-Aldrich (Germany). ABTS (2,2'-azino-bis (3-ethylbenzothiazoline-6-sulphonic acid) diammonium salt) was purchased from Biochem (France). Vanillin was purchased from Merck (Germany). Water was purified by a Milli-Q water purification system.

2.3 Plackett and Burman Design

The Plackett-Burman experimental design was used for screening the effect of six variables which include extraction solution (X_1), Ratio (X_2), extraction mode (X_3), morphotype or cultivar (X_4), variety (X_5) and extraction time (X_6) on the extraction from kola nuts. These six independent variables or factors were organized in eight combinaisons according to Plackett-Burman design matrix [29]. For each factor, a high (+1) and low (-) level were tested (Table 2).

Table 2. Plackett-Burman parameters and coded levels

Factors	Technological parameters	Coded levels	
		Low (-1)	High (+1)
X_1	S: Extraction solution	Water	Ethanol 50%
X_2	R: Ratio (w/v)	1/100	5/100
X_3	M: Extraction mode	Maceration	Infusion
X_4	C: Morphotype or cultivar	White	Red
X_5	V: Variety	Traditional	Improve
X_6	T: Extraction time (h)	3	24

The responses studied were total phenolics content (TPC), total flavonoids content (TFC) and condensed tannin content (CTC). The model created from this analysis of Plackett-Burman experimental design using multiple regression analysis is based on the 1er order-model:

$$Y_n = b_0 + \sum b_i X_i$$

Where Y_n is a predicted response, b_o is a model constant and b_i is a variable linear coefficient.

2.4 Full Factorial Design

A 2^4 full factorial experimental design was used to identify the relationship existing between the response functions and process variables [32], as well as to determine those conditions that optimized extraction of TPC, TFC, and CTC. The four independent variables or factors studied were the extraction solution (X_1), ratio (X_2), extraction mode (X_3) and variety (X_5). Each variable to be optimized was coded at the lower (-1) and higher (+1) levels considered as previously studied (Table 3).

Table 3. Experimental values and code levels of independent variables used for the 2^4 factorial designs

Factors	Technological parameters	Coded levels	
		Low (-1)	High (+1)
X_1	S: extraction solution	Water	Ethanol 50%
X_2	R: ratio (w/v)	1/100	5/100
X_3	M: extraction mode	Maceration	Infusion
X_5	V: variety	Traditional	Improve

In the full factorial design, the main as well as the interaction effects of various factors are determined by fitting the data into 1er order polynomial equation:

$$Y_n = b_0 + b_1 X_1 + b_2 X_2 + ...+ b_k X_k +...+b_{12}X_1X_2 + ... +b_{k-1k}X_{k-1}X_k + ... + b_{1...k}X_1X_2...X_k$$

Where Y_n was the measured response, bk the main effect of the factors X_k, b_{k-1k} the interaction effect between the factors k-1 and k and b_0 the constant term.

2.5 Extraction Procedure

Every dried sample (1 or 5 g) was extracted by diffusion with 100 ml of the extraction solution (water or ethanol 50%) according to the two methods (maceration or infusion) selected. The mixture was least for stand at room temperature during extraction time (3 h or 24 h). Extracts obtained were filtered through a filter paper (Whattman N°1) and stored at 4°C in refrigerator for subsequent determination.

2.6 Analytical Methods

2.6.1 Determination of total polyphenols content (TPC)

Total polyphenols were determined by colorimetry, using the Folin-Ciocalteu method [33,34]. Diluted Folin-Ciocalteu reagent (1/10, v/v, 2.5 mL) was added to 30 μL of sample. After 2 min of incubation in the dark at room temperature, 2 mL of aqueous sodium carbonate (75 g/L) was added. After gentle stirring, the mixture was incubated in a water bath at 50°C for 15 min and rapidly cooled down to stop the reaction. The absorbance was measured at 760 nm with distilled water as blank. A calibration curve was performed with gallic acid at different concentrations (0-1 g/L). Analyses were performed in triplicate and polyphenols level was expressed in milligrams gallic acid equivalent per liter of extract (mg/L GAE).

2.6.2 Determination of total flavonoids content (TFC)

Total flavonoids were determined by the aluminum chloride colorimetric method described by Marinova et al. [35]. In a 25 mL volumetric flask, 0.75 mL of sodium nitrite ($NaNO_2$) distilled water solution (5%, w/v) was added to a 2.5 mL aliquot of the sample. The color reaction was left to develop for 5 min in the dark and at room temperature. Then, 0.75 mL of $AlCl_3$ distilled water solution (10%, w/v) was added and incubated for 6 minutes. After incubation, 5 mL of sodium hydroxide (NaOH 1M) were added and the volume made up to 25 mL. The mixture was mixed well before being dosed with UV-Visible spectrophotometer. The reading was taken at 510 nm with distilled water as a blank. A calibration curve was performed with quercetin at different concentrations (0-1 g/L). The tests were performed in triplicate and the flavonoids content was expressed in milligrams quercetin equivalent per liter of extract (mg/L QE).

2.6.3 Determination of condensed tannins content (CTC)

Condensed tannins were determined by the method of Heilmer et al. [36]. 400 µL of each extract was added to 3 mL of a methanol solution of 4% vanillin and 1.5 mL of concentrated hydrochloric acid was subsquently added. After 15 min of reaction the absorbance was measured at 550 nm with distilled water as blank. A calibration curve was performed with cathechin at different concentrations (0-500 µg/mL). The tests were performed in triplicate and the tannins content was expressed in micrograms catechin equivalent per miligram of extract (mg/L CE).

2.7 Statistical Analysis

All experiments were done in triplicate and data in tables and figures represent mean values ± standard deviation (n=3). Coefficient and experimental standard deviations were determined by the method of linear regression (MS Excel 2007). Comparison of mean values of measured parameters was performed by a one-way ANOVA (STATISTICA, version 7.1) using post hoc Low Statistical Difference (LSD) test. The mean values were considered significantly different when $P=.05$.

3. RESULTS AND DISCUSSION

3.1 Standards Parameters

The absorbance values of stocks solutions of standard gallic acid, quercetin and catechin for Total Polyphenol (TP), Total Flavonoid (TF) and Condensed Tannin (CT), respectively were measured. Table 4 shows the differents equations of each standard linear calibration and

their regression coefficient R^2. It show clearly a good linear relationship between the absorbance and concentration of each standard solution ($0.994 < R^2 < 0.997$). Values gives a good agreement between the experimental and predicted values of the adapted model [37,38].

Table 4. Equation and regression coefficient for different standard calibration

Standards	Equation	R^2
Gallic acid	y=0.872x - 0.006	0.996
Quercetin	y=0.644x - 0.006	0.997
Catechin	y=0.002x - 0.035	0.994

3.2 Screening of Variable Effects on Polyphenols Extraction

According to Plackett-Burman experimental design, 8 experiments were carried out in order to evaluate the effects of the main factors on total phenolics, total flavonoids and condensed tannins extractions. In Table 5, the factors X_1, X_2, X_3, X_4, X_5 and X_6 represent extraction solution, ratio (w/v), extraction mode, morphotype or cultivar, variety and extraction time (h), respectively. The higher level variable was designed as (+1) and the one at the lower level as (-1).

The experimental analysis showed that the yield extraction is favored when the variables extraction solvent, extraction mode, extraction time are in their high level (+1) and the variables ratio, morphotype and variety in their low level (-1) (experiment 6).

Bioactive compounds contents are ranged between 28±1 to 350±13 mg/L GAE, 138±1 to 1653±36 mg/L QE and 33±1 to 365±2 mg/L CE for TPC, TFC and CTC, respectively (Table 5). Total polyphenol, Total flavonoid and condensed tannin yields were influenced by the variables extraction solution (S), ratio (w/v) (R), extraction mode (M) and variety (V).

Table 6 illustrates estimation and statistics of linear regression coefficient. Coefficient is known as statistically significant if its absolute value is strictly higher than the double of the experimental standard deviation, |coef|>2σ [39]. The determination of the each coefficient and experimental standard deviations were determined by the method of linear regression [40].

Table 5. Total polyphenol, total flavonoid and condensed tannin content of kola nuts samples according to Plackett-Burman design

Test set	Independent variables						Experimental responses		
	X_1	X_2	X_3	X_4	X_5	X_6	$Y_1{}^a$	$Y_2{}^b$	$Y_3{}^c$
1	+	+	+	-	+	-	234±3*	777±11	161±1
2	-	+	+	+	-	+	147±5	680±7	205±1
3	-	-	+	+	+	-	113±4**	600±13**	180±1
4	+	-	-	+	+	+	227±9*	1193±22	238±2
5	-	+	-	-	+	+	28±1	138±1	33±1
6	+	-	+	-	-	+	350±13	1653±36	365±2
7	+	+	-	+	-	-	283±2	1157±4	285±1
8	-	-	-	-	-	-	110±7**	617±18**	148±1

Data of the same column having the same sign are statistically in the same homogenous group at P=.05
Y_1: Total polyphenol content; Y_2: Total flavonoid content; Y_3: Condensed tannin content
a: mg/L GAE; b: mg/L QE; c: mg/L CE

Table 6. Statstical estimates of coefficient and standard deviation

	Coefficient and standard deviations for each equation					
	Total polyphenols		Total flavonoids		Condensed tannins	
Coefficient	Values	2σ	Values	2σ	Values	2σ
a_0	186.5*		843.5*		201.94*	
a_1	86.83*		334.83*		60.41*	
a_2	-13.5*		-155.67*		-30.86*	
a_3	24.66*	14	67.33*	53.66	25.79	27.67
a_4	6		64*		25.12	
a_5	-36*		-166.5*		-49.06*	
a_6	1.5		56*		8.40	

*:significant data at P=.05

Statistical analysis showed that extraction solvent, extracton process and variety had the greatest effect on total phenolics extraction. The main variable for condensed tannins extraction were extraction solvent, solid-liquid ratio (w/v) and variety. However, for total flavonoids extraction, all variables were dominating. The levels of the non-dominating factors were fixed in order to allow the optimization.

The factors X_1 (extraction solution), X_2 (solid-liquid ratio), X_3 (extraction mode) and X_5 (variety) have been selected as the most influential in kola nuts compounds extraction according to the effect estimated values (coefficients of the equation). The new experiment was conducted by setting the non-dominating factors "morphotype" (X_4) and "extraction time" (X_6) to their high level (+1), respectively M=Red and T (h)=24.

Minimal effect of the morphotype will make it possible to choose indifferently any cultivar for the future analysis.

3.3 Optimization of Phenolics Compounds Extraction

The full factorial design used was determined the combination of different levels of influential parameters that give the best compounds yields. TFC, TFC and CTC were determined. For that 16 experiments (2^4) were conducted according the matrix presented in Table 7.

The values of regression coefficient determined are given in Table 8. The effect of individual variables and interactions effects was estimated [41].

Table 8 shows that all variables presented significant effect on total phenolics extraction. The most important parameter affecting this extraction is the solvent. The interaction between solvent (X_1) and ratio (X_2) is significant. The predictive equation of total phenolics yields (Y_1), neglecting the non-significant factors, is given by equation 1 with a satisfactory R^2 value (R^2=0.98).

$$Y_1=230.21+49.29X_1-51.04X_2+14.95X_3-31.37X_5+18.87X_1X_2 \qquad (1)$$

The higher total phenolics contents (358 mg/L GAE) was obtained when variables extraction solvent and extraction mode are in their high level (+1) and variables ratio and variety in their low level (-1).

Total flavonoids extraction was affected by solvent (X_1), ratio (X_2) and variety (X_5). No interaction between factors are noted. The most important parameter affecting this extraction is the solvent (Table 8). The data showed a good fit with equation 2, being were statistically acceptable at $P=.05$ level and adequate with a satisfactory R^2 value ($R^2= 0.97$). Equation 2 being developed to present the relationships between TFC and extraction variables.

$$Y_2=854.08+325.92X_1-119.25X_2-149.58X_5 \quad (2)$$

The highest value of total flavonoid (1449 mg/L QE) was obtained when the variable extraction solvent is in his high level (+1) and variables ratio and variety in their low level (-1).

Condensed tannin extraction was affected by by all variables. The most important parameter affecting tannin extraction is the same as in the case of total phenolics extraction (factor X_1). Two significants interactions were observed: solvent (X_1) - ratio (X_2) and solvent (X_1) – variety (X_5). Equation 3 describe the model of condensed tannin with a satisfactory R^2 value ($R^2= 0.96$).

$$Y_3=159.85+58.84X_1+22.73X_2+21.06X_3-$$
$$20.53X_5+23.24X_1X_2+15.62X_1X_5 \quad (3)$$

The highest value of condensed tannin (267.39 mg/L CE) was obtained also when variables solvent, ratio, extraction mode are in their high level and variable variety in their low level.

The experimental analysis showed that polyphenolics compounds extraction of kola nuts is favored when the variables extraction solvent and mode are in their high level (+1), and variables ratio and variety in their low level (-1). Thus, the optimum extraction process of kola nuts bioactive compounds involves the following parameters:

- Extraction solvent : Aqueous ethanol (50%)

Selecting the right solvent affects the amount and rate of polyphenols extracted [42]. The use of water with a organic solvent, for the extraction of polyphenols from some plant materials has been reported to contribute to the creation of a moderately polar medium that insures the extraction of phenolics, giving betters results than when using a pure organic solvent [12,13]. According to Mokhtarpour et al. [15], it was suggested the using of 50% aqueous ethanol for plant tannin extraction. Ethanol is a good solvent for polyphenol extraction and is safe for human consumption [43].

- Ratio (w/v) : 1/100

A correct ratio of solvent and plant matter is fundamental for obtaining an optimal extraction process. We note an increasing of TPC, TFC and CTC when extraction solution volume move from 20 mL (5/100) to 100 mL (1/100). According to Sampath [10], when the solvent volume was increased, it can increase the absorption rate, swelling rate and diffusion rate of the plant cell wall. At the same time, excessive solvent volume, promotes the extraction of undesired compound from the plant material, affect the quality of desired compounds and decrease the yield also. This decrease due to the fact that when the ratio reached a certain level, the extract may be well saturated [44,45,38].

- Extraction mode : infusion

The temperature increase would favor the diffusion of the compounds of the vegetable matrix to the extraction solution [12]. In this study, heating the extraction solvent promotes the diffusion of the sample compounds, kola nuts bioactive compounds would thermostable. Temperature's effect on extraction is dual. On one hand, higher temperature can accelerate the solvent flow and thus increase the content and on the other hand, higher temperature can decrease the fluid density that may reduce the extraction efficiency [45].

- Variety : traditional

Genetic modification of kola nut could explain the higher efficiency of traditional nuts compared to improved nuts. A similar report by Nyamien et al. [38], revealed that kola nuts caffeine content depends on the type of variety used.

3.4 Validation of 2^4 Full Factorial Design Optimization of Phenolics Compounds from Kola Nuts

All the models were established with high coefficient of determination R^2, ranging from 0.96 to 0.98, wich means a close agreement between the experimental results and those predicted by the models. The predictive quality of every model was also tested at the recommended optimum condition. All the responses were replicated three times at the optimum condition, and the results are presented in Table 9. The arithmetic means of the experimental values was 350±11 mg/L GAE, 1460±9 mg/L QE and 264.33±2 mg/L CE for TPC, TFC and CTC, respectively.

Table 7. Experimental design (2^k, k=4) and corresponding responses

Run order	Technological parameters and coded levels				Responses		
	X_1	X_2	X_3	X_5	$Y_1{}^a$	$Y_2{}^b$	$Y_3{}^c$
1	W	1/100	M	T	247±18**	727±4	120±1
2	E	1/100	M	T	347±4+	1587±16	203±1
3	W	5/100	M	T	119±4	459±11*	97±1**
4	E	5/100	M	T	288±13***	1253±11	282±1
5	W	1/100	I	T	313±9	933±4	159±1
6	E	1/100	I	T	353±9+	1373±4	190±1
7	W	5/100	I	T	167±22	571±9	173±7
8	E	5/100	I	T	259±31**	1127±4**	241±1
9	W	1/100	M	Ip	193±11*	440±1*	30±1*
10	E	1/100	M	Ip	263±11**	1013±16	139±1
11	W	5/100	M	Ip	67±9	187±9	31±1*
12	E	5/100	M	Ip	197±4*	873±9	214±1
13	W	1/100	I	Ip	250±1**	620±1	97±1**
14	E	1/100	I	Ip	283±11***	1093±4**	138±1
15	W	5/100	I	Ip	91±4	289±11	100±1
16	E	5/100	I	Ip	245±11**	1120±13**	327±1

Data of the same column having the same sign are statistically in the same homogenous group at P=.05
W: water ; E: ethanol 50% ; I: infusion ; M: maceration ; T: traditional ; Ip: improve a: mg/L GAE; b: mg/L QE; c: mg/L CE

Table 8. Statstical estimates of coefficient and standard deviation of 2^4 full factorial experiments

Coef	Model coefficients					
	Total polyphenols		Total flavonoids		Condensed tannins	
	Values	2σ	Values	2σ	Values	2σ
b_0	230.21*		854.08*		159.45*	
b_1	49.29*		325.92*		58.84*	
b_2	-51.04*		-119.25*		22.73*	
b_3	14.95*		36.75		21.06*	
b_5	-31.37*		-149.58*		-20.53*	
b_{12}	18.87*	9.66	32.58	51.98	23.24*	13.05
b_{13}	-9.29		-38.41		-10.52	
b_{15}	-0.79		-5.41		15.62*	
b_{23}	-3.79		5.08		6.8	
b_{25}	2.37		32.08		10.26	
b_{35}	3.54		39.42		10.58	

: significant data, Coefficient is known as statistically significant if its absolute value is strictly higher than the double of the experimental standard deviation, |coef|>2σ [39]

Table 9. Experimental data for verification of the models predicted at optimal condition

Optimal condition	Response 1 TPC (mg/L GAE)		Response 2 TFC (mg/L QE)		Response 3 CTC (mg/L CE)	
	Pred	Exp	Pred	Exp	Pred	Exp
X_1=Ethanol 50% X_2= Ratio 1/100 X_3= Infusion X_4= Traditional	358*	350±11*	1449**	1460±9**	267.39***	264.33±2***

Data of the same column having the same sign are statistically in the same homogenous group at P=.05
Pred= predicted values, Exp= experimental values

Experimented data were approaching the predicted values. This indicated that the optimization achieved in the preent study was reliable. Deviations between experimental values and the predicted values can be explained by the lack of perfectly fitted models and experimental errors.

4. CONCLUSION

Kola nuts occupy a unique place amongst West Africans where these are widely consumed. These are the important source of bioactive compounds, and can be used as a possible source of natural antioxidants for African's population and European industries. The use of experimental design in this study showed the optimum conditions for extracting bioactive compound from kola nuts by non toxic solvent (aqueous ethanol in particular). All the developed models showed an adequate predictive quality and the optimal condition was predicted by desirability function and successively verified. The optimum conditions for obtaining high yield of polyphenols content were extraction by infusion of traditional variety at solid-liquid ratio 1/100 with aqueous ethanol 50%. A meticulous study of each factor must be done to obtain more significant yields of kola nuts compounds. The researchers may determine precise phenolics contents of the most widely eaten kola to avoid consumption in high dose since our knowledge on risks is much limited. Extraction with non toxic solvent may allow the isolation of directly edible ingredients.

COMPETING INTERESTS

Authors have declared that no competing interests exist.

REFERENCES

1. Momo C, Ngwa A, Dongmo G, Oben J. Antioxidant properties and alpha-amylase inhibition of *Terminalia superba*, *Albizia* sp., *Cola nitida*, *Cola odorata* and *Harungana madagascarensis* used in the management of diabetes in Cameroon. J. Health Sci. 2009;55(5):732-738.
2. Laib I and Barkat M. Composition chimique et activité antioxydante de l'huile essentielle des fleurs sèches de *Lavandula officinalis*. Agric. 2011;2:89-101. French
3. Fernande N, Marthe T, Laure N, Enyong J. *In vitro* antioxidant activity of *Guibourtia tessmannii* Harms, J. Leonard (Cesalpinoidae). J. Med. Plant Res. 2013; 7(42):3081-3088.
4. Mosquera O, Correa Y, Buitrago D, Nino J. Antioxidant activity of twenty five plants from Colombian biodiversity. Mem Inst Oswaldo Cruz. 2007;102(5):631-634.
5. Scalbert A, Manach C, Morand C, Rémésy C. Dietary polyphenols and the prevention of diseases. Crit Rev Food Sci Nutr. 2005; 45:287-306.
6. Omar A. The antioxidant activity and polyphenolic contents of different plant seeds extracts. Pak. J. Biol. Sci. 2009; 12(15):1063-1068.
7. Santos D, Cavalcanti R, Rostagno M, Queiroga C, Eberlin M, Meireles A. Extraction of polyphenols and anthocyanins from the Jambul (*Syzygium cumini*) fruits peels. Food Public Health. 2013;3(1):12-20.
8. Dai J, Mumper R. Plant phenolics: Extraction, analysis and their antioxidant and anticancer properties. Molecules. 2010;15:7313-7352.
9. Tatiya A, Tapadiya G, Kotecha S, Surana S. Effect of solvents on total phenolics, antioxidant and antimicrobial properties of *Bridelia retusa* Spreng. stem bark. Indian J. Nat. Prod. Resour. 2011;2(4):442-447.
10. Sampath M. Optimization of the extraction process of phenolic antioxidant from *Polyalthia longifolia* (Sonn.) Thawaites. J. Appl. Pharm. Sci. 2013;3(2):148-152.
11. Cheng A, Yan H, Han C, Chen X, Wang W, Xie C, et al. Acid and alkaline hydrolysis extraction of non-extractable polyphenols in blueberries: optimisation by response surface methodology. Czech J. Food Sci. 2014;32(3):218-225.
12. Shi J, Yu J, Pohorly J, Young C, Bryan M, Wu Y. Optimization of the extraction of polyphenols from grape seed meal by aqueous ethanol solution. Food Agric. Environ. 2003;1(2):42-47.
13. Mussatto S, Ballesteros L, Martins S, Teixeira J. Extraction of antioxidant phenolic compounds from spent coffee grounds. Sep. Purif. Technol. 2011;83:173-179.
14. Selvaraj K, Chowdhury R, Bhattacharjee C. Optimization of the solvent extraction of bioactive polyphenolic compounds from aquatic fern *Azolla microphylla* using response surface methodology. Int. Food Res. J. 2014;21(4):1559-1567.

15. Mokhtarpour A, Naserian A, Valizadeh R, Mesgaran D, Pourmollae F. Extraction of phenolic compounds and tannins from Pistachio by-products. Annu. Res. Rev. Biol. 2014;4(8):1330-1338.

16. Koffi E, Sea T, Dodeye Y, Soro S. Effect of solvent type on extraction of polyphenols from twenty three Ivorian plants. Journal of Animal and Plant Sciences. 2010;5(3):550-558.

17. Arhewoh M, Falodun A, Okhamade A, Boa Y, Sheng Q. Ultrasonic assisted extraction and radical scavenging activity of some selected medicinal plants. J. Pharm. Res. 2011;4(2):408-410.

18. Anurukvorakun O. Factorial design applied to subcritical water extraction for the investigation of flavonoids and antioxidant capacity of *Gynura calciphila* Kerr. Mahidol University J. Pharma. Sci. 2013;40(2):7-16.

19. Nyamien Y, Adjé F, Niamké F, Chatigre O, Adima A, Biego H. Caffeine and phenolic compounds in *Cola nitida* (Vent.) Schott and Endl and *Garcinia kola* Heckel grown in Côte d'Ivoire. Br. J. Appl. Sci. Tech. 2014;4(35):4846-4859.

20. Kuzma P, Druzynska B, Obiedzinski M. Optimization of extraction conditions of some polyphenolic compounds from parsley leaves (*Petroselinum crispum*). Acta Sci. Pol., Technol. Aliment. 2014; 13(2):145-154.

21. Niemenak N, Onomo P, Fotso, Lieberei R, Ndoumou D. Purine alkaloids and phenolic compounds in three *Cola* species and *Garcinia kola* grown in Cameroon. S. Afr. J. Bot. 2008;74:629-638.

22. Boudjeko T, Rihouey C, Ndoumou D, El Hadrami I. Characterisation of cell wall polysaccharides, arabinogalactans-proteins (AGPs) and phenolics of *Cola nitida*, *Cola acuminata* and *Garcinia kola* seeds. Carbohydr. Polym. 2009;78:820-827.

23. Tende A, Ezekiel I, Dare S, Okpanachi O, Kemuma O, Goji T. Study of the effect of aqueous extract of kolanut (*Cola nitida*) on gastric acid secretion and ulcer in the white wistar rats. Br. J. Pharmacol. Toxicol. 2011;2(3):132-134.

24. Aikpokpodion P, Oduwole O, Adebiyi S. Apparaisal of pesticide residues in kola nuts obtained from selected markets in Southwestern, Nigeria. J Sci. Res. Rep. 2013;2(2):582-597.

25. Prohp T, Ekpo K, Osagie E, Osagie A, Obi H. Polyphenol contents and polyphenol oxidase activities of some Nigerian kolanuts. Pak. J. Nutr. 2009;8(7):1030-1031.

26. Okoli J, Abdullahi K, Myina O, Iwu G. Caffeine content of three Nigerian cola. J. Emerg. Trends Eng. Appl. Sci. 2012;3(5):830-833.

27. Arhewoh M, Falodun A, Okhamade A, Boa Y, Sheng Q. Ultrasonic assisted extraction and radical scavenging activity of some selected medicinal plants. J. Pharm. Res. 2011;4(2):408-410.

28. Umoren E, Osimand E, Udoh P. The comparative effects of chronic consumption of kola nut (*Cola nitida*) and caffeine diets on locomotor behaviour and body weights in mice. Niger. J.Physiol. Sci. 2009;24(1):73-78.

29. Rajendran A, Thirugnanam M, Thangavelu V. Statistical evaluation of medium components by Plackett-Burmann experimental design and kinetic modeling of lipase production by Pseudomonas fluorescens. Indian J. Biotechnol. 2007; 6:469-478.

30. Analytical Methods Committee, AMCTB N°55. Experimental design and optimization (4): Plackett-Burman designs. Anal. Methods. 2013;5:1901-1903.

31. Wahidu Z. Optimization of antioxidant extraction from jackfruit (*Artocarpus heterophyllus* Lam.) seeds using response surface methodology. Master of Science in Nutrition and Natural Development, Faculty of Bioscience Engineering: Ghent University; 2012.

32. Aboua N, Yao B, Gueu S, Trokourey A. Optimization by experimental design of activated carbons preparation and their use for lead and chromium ions sorption. Res. J. Agric. Biol. Sci. 2010;6(6):665-670.

33. Mylonaki S, Kiassos E, Makris D, Kefalas P. Optimization of the extraction of olive (*Olea europea*) leaf phenolics using water/ethanol based solvent system and response surface methodology, Anal. Bioanal. Chem. 2008;392(5):977–985.

34. Singleton V, Rossi J. Colorimetry of total phenolics with phosphomolybdic-phosphotungstic acid reagents. Am. J. Enol. Viticult. 1965; 16:144-158.

35. Koffi E, Le Guernevé C, Lozano P, Meudec E, Adjé F, Bekro Y et al. Polyphenol extraction and characterization of *Justicia secunda* Vahl leaves for traditional medicinal uses. Ind Crpos Prod. 2013;49:682-689.

36. Marinova D, Ribarova F, Atanassova M. Total phenolics in bulgarian fruits and vegetables. J. Univ. Chemi. Technol. Metal. 2005;40(3):255-260.

37. Heimler D, Vignolini P, Din M, Vinueri F, Ronani A. Antiradical activity and polyphenol composition of local Brassiccaceae edible varieties. Food Chem. 2006;99(3):464-469.

38. Mouchine F, Abdellah F, Bouchaib I, Taoufik H, Saad R. The application of Plackett and Burman design in screening the parameters acting on the hydrodistillation process of Moroccan rosemary (*Rosmarinus officinalis* L.). Int. J. Innov. Appl. Stud. 2014;8(1):372-381.

39. Nyamien Y, Adjé F, Niamké F, Koffi E, Chatigre O, Adima A, et al. Effect of solvents and solid-liquid ratio on caffeine extraction from Côte d'Ivoire kola nuts (*Cola nitida*). Int. J. Sci. Res. 2015;4(1):218-222.

40. Assidjo E, Yao B, Akou E, Ado G. Optimisation of the treatment conditions of cocoa butter in order to reduce non-quality. Journal of chemom. 2005;19:543-548.

41. Feinberg M. La validation des méthodes d'analyse: Approche chimiométrique de l'assurance qualité au laboratoire, Paris:Masson; 1996. French.

42. Nese Ö, Duygu K. Boron removal from aqueous solutions by batch absorption onto cerium oxide using full factorial design. Desalination. 2008;223(1):106-112.

43. Xu B and Chang S. A comparative study on phenolic profiles and antioxidant activities of legumes as affected by extraction solvents. J Food Sci. 2007;72(2):159-166.

44. Xi J. Caffeine extraction from green tea leaves assisted by high pressure processing. J. Food Eng. 2009;94:105-109.

45. Sathishkumar T, Shanmugam S, Rajasaekaran P, Sadavivam S, Manikandan V. Optimization of flavonoids extraction from the leaves of *Tabernaemontana heynena* Wall. using L_{16} orthogonal design. Nat. Sci. 2008;6(3):10-21.

11

Fatty Acid Composition and Physicochemical Properties of Four Varieties of *Citrullus lanatus* Seed Oils Cultivated in Côte d'Ivoire

Anne Marie Niangoran N'guetta[1], Yolande Dogoré Digbeu[2], Siaka Binaté[1], Jean Parfait Eugène N'guessan Kouadio[1], Soumaila Dabonné[1] and Edmond Ahipo Dué[1*]

[1]*Biochemistry and Food Technology Laboratory, Nangui Abrogoua University, 02 BP 801 Abidjan 02, Côte d'Ivoire.*
[2]*Laboratory of Food Safety, Lorougnon Guede University, BP 150 Daloa, Côte d'Ivoire.*

Authors' contributions

This work was carried out in collaboration between all authors. Author EAD designed the study, wrote the protocol and wrote the first draft of the manuscript. Authors AMNN carried out all laboratoire work. Authors SB and YDD performed the statistical analysis, managed the analyses of the study. Authors SD and JPENK managed the literature searches. All authors read and approved the final manuscript.

Editor(s):
(1) Miguel Cerqueira, Institute for Biotechnology and Bioengineering, Centre of Biological Engineering, University of Minho, Portugal.
Reviewers:
(1) Emmanuel Ibitola Bello, Department of Mechanical Engineering, The Federal University of Technology, Akure, Nigeria.
(2) I.O. Olabanji, Chemistry, OAU, Ile-Ife; Nigeria.
(3) Zoue Lessoy Yves Thierry, University Felix Houphouet Boigny, Abidjan, Côte d'Ivoire.
(4) Anonymous, India.

ABSTRACT

In this study, the physicochemical properties and fatty acids composition of four varieties of *Citrullus lanatus seed* oils cultivated in Côte d'Ivoire were investigated. There was one variety (Bebu cultivar) with oval flat seeds and thick edges (COS), and three varieties (wlêwlê cultivars) with flat seeds disentangled and sharp extremity of various size (big seed (CBS), average seed (CAS) and small seed (CSS)). The pH at 30ºC was ranged from 5.85±0.003-6.03±0.001. The results for density (g/ml), refractive index, unsaponifiable matter (% of oil) and impurity (%) were 0.876±0.005-0.924±0.001; 1.4595±0.004-1.4739±0.001; 0.89±0.450-2.07±0.160; 0.05±0.002-

Corresponding author: Email: ahipoedmond@yahoo.fr

0.1±0.001; respectively. The free fatty acid (%), acid value (mgKOH/g oil), peroxide index (meq O2/kg oil), saponification value (mgKOH/g oil), iodine value (mg of g I2/100 g oil), Total polyphénolics content (mg gallic acid/kg oil) values of the oils were obtained in the range; 0.80±0.050-2.03±0.010; 6.03±0.350-7.93±0.100; 10.93±0.470-13.80±0.100; 228.43±0.040-244.76±0.010; 84.07±11.250-93.38±1.090; 0.125±0.050-0.291±0.010; respectively. The most abundant fatty acid detected in the four oils was oleic acid ranged from 46.56-59.97. This acid constituted with the aldaic acid, the monounsaturated fatty acids in these oils and accounted for 64.82-73.07. This study showed the presence of others main saturated fatty acids as stearic acid and palmitic acid. The properties of these oils were quite comparable to those of previously reported cucurbitaceaes seeds oils, suggesting their potential use as good table and cooking oils. With a high yield of oil and physicochemical characteristics similar to those of the other commercial edible oils, the Citrullus lanatus seed oil can be considered as a new and valuable source of edible oil. This study also showed potential for Citrullus lanatus seeds oils to ave relatively high oxidative stability that would be suitable for food and industrial applications.

Keywords: Citrullus lanatus; varieties; seed oil; physicochemical properties; fatty acids composition.

1. INTRODUCTION

Citrillus lanatus belongs to the family of Cucurbitaceae which has a tremendous genetic diversity, extending to vegetative and reproductive characteristics [1]. They thrive in tropical, subtropical, arid deserts and temperate locations. This specie is an annual, herbaceous, monoecious plant with a non-climbing creeping habit [1]. This flowering plant produces a special type of fruit known by botanists as a pepo, which has a thick rind (exocarp) and fleshy centre (mesocarp and endocarp); pepos are derived from an inferior ovary and are characteristic of the Cucurbitaceae. The watermelon fruit, loosely considered a type of melon has a smooth exterior rind (green and yellow) and a juicy, sweet, usually red, but sometimes orange, yellow, or pink interor flesh [2]. Citrullus lanatus seeds have both nutritional and cosmetic importance, the seed contain vitamins (B1, minerals, riboflavin, fat, carbohydrates and proteins [3].

In west Africa, a region where soups are integral to life, Citrillus lanatus seeds are a major soup ingredient and a common component of daily meals. Coarsely ground up, they thicken stews and contribute to widely enjoyed steam dumplings. Some are soaked, fermented, boiled and wrapped in leaves to form a favourite food seasoning. Despite being a significant foodstuff even by global standard, pistache is hardly known to nutritionists outside a few west African nations. Little nutrional detail on Citrullus lanatus seed oil is readily available to an international readership. Research studies have shown that these seeds contained about 50% oil [4], 42-57% oil [5], 44-53% oil [6] for seeds cultivated in

different bioclimatic regions of cameroon and 56,9% oil for Citrullus lanatus from Nigeria [7]. These studies showed that Citrullus lanatus seeds contained good amounts of oil that can be exploited. The aim of the present study therefore is to determine some physical and chemical properties as well as the fatty acid composition from four varieties of Citrullus lanatus seeds oils cultivated in Côte d'Ivoire.

2. MATERIALS AND METHODS

2.1 Sample Collection

This study was carried out on four varieties of Citrullus lanatus grown at the experimental station of Nangui Abrogoua University, at Abidjan Côte d'Ivoire. These varieties were authenticated by a taxonomist in the Department of crop Science of this University. At maturity (3-4 months), the fruits were harvested and transported to laboratory. After the berries fermentation (5 days at 25°C), seeds of the four obtained varieties were washed, dried (oven drying (60°C) during 24 hours) and peeled. For this work, one variety (Bebu cultivar) with oval flat seeds and thick edges (COS) and three varieties (wlêwlê cultivars) with flat seeds disentangled and sharp extremity of various size (big seed (CBS), average seed CAS) and small seed (CSS)) were used (Fig. 1).

2.2 Oil Extraction

The dried samples (200g) were ground into powder with the laboratory pestle and mortar and kept or stored in air tight containers until required for subsequent analysis. Oil from seeds was extracted in soxhlet using hexane as solvent and the lipid value (crude fat) was evaluated. The oil extracted after drying was transferred into a translucent glass bottle and kept in the refrigerator (4°C) until required for analysis. The peeled seeds are dried in the steam room at 40°C during 24 hours. 200g of almonds were weighed and crushed by means of a mixer. The obtained powder (the taking), was used to extract the oil. Extraction was done for heating during 12 hours. The organic phase containing the fat is evaporated until elimination of the solvent. Then, balloon containing oil is placed in the steam room at 70°C during 30 min for the evaporation of solvent.

2.3 Physicochemical Properties Analysis

The physicochemical parameters (refractive index, acid, iodine, peroxide and saponification values, density) were carried out according to the methods described [8]. Colour was measured with a Lovibond tintometer ; unsaponifiable matters was carried out by the procedure of Lozano et al. [9]. Determination of the total polyphenolics content was done spectrophotometrically using Folin-Ciocalteau's reagent as described by Capannesi et al. [10]. The determination of the pH value is carried out by measurement of the potential difference between electrodes immersed in standard and test solutions. The standard solutions used are assigned a definite pH value by convention.

2.4 Fatty Acid Composition Analysis

The fatty acid composition of oils was determined by gas chromatography (GC) as fatty acid methyl esters (FAMEs). FAMEs were prepared by saponification/methylation with sodium methylate according to European Regulations (EEC 2568/91). A chromatographic analysis was performed in a Hewlett-Packard model 4890D gas chromatograph equipped with a 30m x 0.25 mm x 0.25µm film thickness fused silica capillary column (Innowax) coupled to a flame ionization detector (column temperature 210°C). Both the injector and the detector were maintained at 230°C and 250°C, respectively. Nitrogen was used as the carrier gas at 1ml/min with Split injector system (Split ratio 1 :100). Fatty acids were identified by comparing their retention times with those of standard compounds (butyric; caprylic; capric; lauric; mirystoleic; myristic; palmitoleic; palmitic; linolenic; linoleic; oleic; aldaric; stearic; arachibonic and arachidique acids from Sigma-Aldrich).

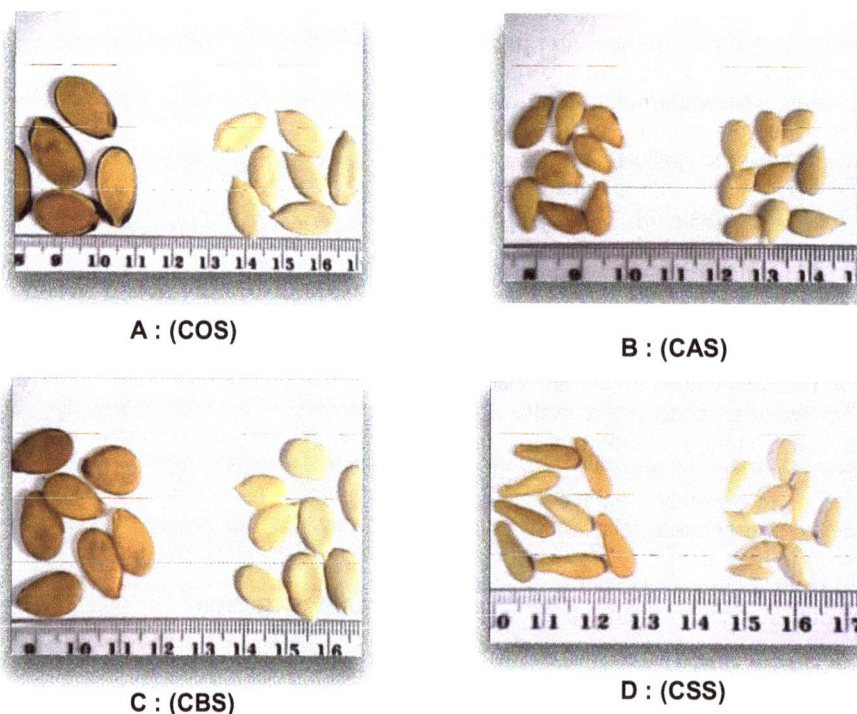

A : (COS)

B : (CAS)

C : (CBS)

D : (CSS)

Fig. 1. Seeds of cucurbit *Citrullus lanatus* varieties: A: Oval flat seeds and thick edges (COS); B: average seeds (CAS); C: big seeds (CBS); D: small seeds (CSS)

3. RESULTS AND DISCUSSION

3.1 Physicochemical Characterization of Four Varieties of *Citrillus lanatus* Seed Oils

The total lipid content of *Citrullus lanatus* seeds (Table 1) were 56.08±2.16; 55.62±0.49; 51.43±1.05; 49.24±0.77 respectively for CSS; CAS; CBS and COS. From the results, these varieties of *Citrullus lanatus* could be considered as good lipid sources. These values were similar to those reported by Essien and Eduok [11] for southern Nigeria *Citrullus lanatus* seed oil (57.26 %), Fokou et al. [12], for Cucurbitaceae oils from four different regions in Cameroon, (49.01-52.15g/100g) but higher in value than L. breviflora seed oil (22.9%) [13] or *Lagenaria siceraria* (Mol.) cultivars from Nigeria (38.10-43.65%) [14].

Physicochemical properties of ivorian *Citrullus lanatus* seeds oils are shown in Table 1.

The refractive indices measured in the present work (ranging from 1.4595-1.4739), were similar to those reported by Lazos [3] for pumpkin (1.4616), Oluba et al. [7] for egusi melon seed oil (1.45) or Ardabili et al. [15] for pumpkin seeds (1.4662).

This shows that these oils are less thick compared with most drying oils whose refractive indices were between 1.48 and 1.49 [16].

The refractive index is used by most processors to measure the range in unsaturation as the fat or oil is hydrogenated. The refractive index of oils depend on their molecular weight, fatty acid chain length, degree of unsaturation [17].

Saponification index is an indicator of the average molecular weight and hence, chain length. It depends on the neutralization of the free fatty acids in oil and complete saponification of the fatty material [18]. The saponification indices of Ivorian *citrullus lanatus* (CSS; CAS; CBS and COS) were respectively 244.76; 228.43; 239.61 and 239.34 mg KOH/g of oil. These values agree with those of saturated fatty acids rich oils such as *Cocos nucifera* coconut (248-265 mg KOH/g of oil) and *Elais guineensis*, palm kernel oils (230-254 mg KOH/g of oil) [19]. As reported by Pearson [20], oil with higher saponification values contain high proportion of lower fatty acids. Therefore, the values obtained for *citrullus lanatus* oils in this study show that they contain high amounts of higher fatty acids.

These values were also slightly higher than those of non-conventional oils such as Dacryodes elulis, the African pear (201.4 mg KOH/g of oil) [21], Coula edulis (180-185 mg KOH/g of oil) [22], C. schweinfurthii (177-197.79 mg KOH/g of oil) [23].

The iodine indices obtained in this study for the studied oils were 93.38; 92.97 83.25 84.07±11.25, g I$_2$/100g respectively for CSS; CAS; CBS and COS. The iodine index is a characteristic of the unsaturation of a fatty acids or its esters. Lipids with unsaturated fatty acids (containing one or more double bonds) are easily assimilated and broken down to produce calorific energy than saturated fatty acids. The higher the iodine value, the more unsaturated the oil. However, when the iodine value becomes too high, the stability of the oil reduces because it is more likely to undergo oxidation. These iodine values of CSS, CAS and CBS oils, ranged (80 to 100 g I$_2$/100g), were similar to those of unsaturated fatty acid rich oils such as peanut (86.06-107.0 g I$_2$/100g), cotton seed (100.0-123.0 g I$_2$/100g), but lower than that of sunflower (118.0-141.0 g I$_2$/100g), of soybean oil (124.0-139.0 g I$_2$/100g) [24]. However, C. lanatus has higher iodine value than those of saturated fatty acid-rich oils such as Theobroma cacao, cocoa butter (32.0-42.0 g I$_2$/100g) [25], coconut (6.0-10.0 g I$_2$/100g), palm oil (50.0-55.0 g I$_2$/100g), palm kernel (14.0-1.0 g I$_2$/100g) [24].

The percentage of impurity (acid/saponification index) ranged from 0.01 (COS) to 0.05 (CSS; CAS and CBS). The percentage of impurities (high amount of free fatty acids) may be due to processing methods, storage conditions and extraction method used in the laboratory [25].

Unsaponifiable matters in the vegetale oils are a variety of nonglyceridic bioactive substances containing variable mixture of hydrocarbons, aldehydes, ketones, alcohols, sterols, pigments, and fat-soluble vitamins that may occur naturally or may be formed during processing or degradation of oils [26]. The content of unsaponifiable matters (2.07±0.16; 0.89±0.45; 1.31±0.32; 1.31±0.31) for CSS; CAS; CBS and COS respectively were much lower than the values reported by Answar et al. [27] for rice bran oil (3-7%), Ardabili et al. [15], for the pumpkin seed oil (5.7±0.8%). But *citrullus lanatus* unsaponifiable matters from CAS; CBS and COS were in a close agreement with the result of Milovanovic and Jovanovic [28], for *Citrullus colocynthis* L. seed oil (1.02%).

The acid value represents free fatty acid content due to enzymatic activity. The acid values were 7.93±0.1; 6.92±0.17; 6.03±0.35 and 7.23±0.13 mg KOH/g for CSS; CAS; CBS and COS respectively. While the free fatty acid values were 1.84±0.02; 2.03±0.01; 1.53±0.03; 0.80±0.05% respectively which were below 5.00% free fatty acids content recommended as the maximum for non-rancid oil [29,30]. This implies that the studied oils were not rancid.

The phenolic compounds obtained in this study were 0.125±0.05; 0.291±0.01; 0.251±0.013 and 0.167±0.03 g/l respectively for *citrullus lanatus* CSS; CAS; CBS and COS. These compounds represent potentially health-promoting substances and have industrial applications [31]. These naturally occuring compounds have proven to play important role in the stability and sensory and nutritional characteristics of the product and may prevent deterioration through quenching of radical reactions responsible for lipid oxidation [32]. Phenolic compounds values of this work were in agreement with those usually obtained from seed oils (0.1 to 0.3 mg/kg) [15]. Cultivar, extraction system, and the conditions of processing and storage are critical factors for the content of phenolic compounds [33].

3.2 Oil Fatty Acid Composition

Table 2 shows that the most abundant fatty acid detected in the four varieties of *Citrullus lanatus* seeds oils from Côte d'Ivoire was oleic acid (46.56-59.97).

This study showed the presence of others main fatty acids as elaidic acid (13.10-21.23), palmitic acid (12.03-13.83) and stearic acid (09.49-11.84) in the oils.

The monounsaturated fatty acids (MUFA) of cultivars oils we studied (64.82; 69.90 ; 73.07, 67.79, respectively for CBS, COS, CAS, CSS) were similar to those reported by Essien et *al.* [14] for *lagenaria siceraria* seeds oils (66.68-71.89%) but higher than those reported by Oluba et al. [7], for Nigerian *Citrullus lanatus* seeds oils (57.4%). As can be seen in Table 2, monounsaturated fatty acid have great importance because of their nutritional implication and effect on the oxidative stability oils. These oils high in MUFA have been well documented to provide numerous health benefits [34,35]. Our results with oleic acid as the predominant level in these *Citrullus lanatus* seeds oils were so similar to those of Idouraine et al. [36] and Zdunczyk et al. [37] who showed that C. pepo seeds contain oleic acid as the most abundant fatty acid and C. pepo subsp. These results revealed that ivorian's *Citrullus lanatus* seeds oils were in agreement with all other studies of Cucurbitaceae seeds oils which were low in linolenic acid (<1%) ([38-40]). Unsaturated fatty acids values of *Citrillus lanatus* seeds oils from Cote d'Ivoire suggest that they could be some goods sources of edibles oils (for cooking) and a potential antidote for fight against cardiovascular disease.

Table 1. Physicochemical characteristics of four varieties of *Citrullus lanatus* seeds oils from Côte d'Ivoire

Cultivars \ Parameters	CSS	CAS	CBS	COS
Total lipid content	56.08±2.16[a]	55.62±0.49[a]	51.43±1.05[a]	49.24±0.77[a]
pH at 30°C	5.93±0.001[b]	5.90±0.002[b]	5.85±0.003[c]	6.03±0.001[a]
Density (g/ml)	0.891±0.003[b]	0.924±0.001[a]	0.916±0.001[a]	0.876±0.005[c]
Refractive Index	1.4732±0.001[a]	1.4686±0.003[a]	1.4739±0.001[a]	1.4595±0.004[a]
Unsaponifiable matter (% of oil)	2.07±0.16[a]	0.89±0.45[c]	1.31±0.32[b]	1.31±0.31[b]
Impurity (%)	0.05±0.002[b]	0.05±0.002[b]	0.05±0.002[b]	0.1±0.001[a]
Free fatty acid, (% of oleic acid)	1.84±0.02[b]	2.03±0.01[a]	1.53±0.03[c]	0.80±0.05[d]
Acid value (mgKOH/g of oil)	7.93±0.1[a]	6,92±0.17[b]	6.03±0.35[c]	7.23±0.13[b]
Peroxide index meq O2/kg of oil	12.34±0.23[b]	10.93±0.47[c]	12.38±0.3[b]	13.80±0.1[a]
Saponification value mgKOH/g oil	244.76±0.01[a]	228.43±0.04[c]	239.61±0.02[b]	239.34±0.02[b]
Iodine value (g I_2/100 g of oil)	93.38±1.09[a]	92.97±0.98[a]	83.25±1.34[b]	84.07±11.25[b]
Total polyphénolics content (mg gallic acid/kg oil)	0.125±0.05[c]	0.291±0.01[a]	0.251±0.013[b]	0.167±0.03[c]
Color	Yellow orange	yellow	Light yellow	Light yellow

Mean± standard deviation of three determinations. In the same row, mean values followed by the same letter (superscript) are not significantly different (p>0.05)

Table 2. Fatty acids composition of four varieties of *Citrullus lanatus* seeds oils from Côte d'Ivoire

samples Fatty acid	Carbons and bonds	CBS	COS	CAS	CSS
Lauric acid	C12 :0	1.67	1.07	1.03	1.34
Myristic acid	C14 :0	1.10	_	_	1.24
Palmitic acid	C16 :0	13.82	13.44	12.03	12.71
Stearic acid	C18 :0	11.84	11.62	09.49	09.71
Oleic acid	C18 :1	50.27	52.68	59.97	46.56
Elaidic acid	C18 :1 Trans-9	14.55	17.22	13.10	21.23
Saturated fatty acids (SFA)		28.43	26.13	22.55	25
Monounsaturated fatty acids (MUFA)		64.82	69.90	73.07	67.79

The four cucurbits oils studied contents low amounts of saturated fatty acids (22.55-28.43) and high contents of unsaturated fatty acids (64.82-73.07). It is now widely accepted that diet with low saturated fatty acids and high in unsaturated fatty acids is beneficial for health.

The SFA such as lauric; myristic and to a lesser extent, palmitic acids, elevate serum cholesterol and LDL levels. Stearic acid does not elevate serum cholesterol or LDL levels, but its other health effects are yet undefined [41].

The colour of the oil is used preliminarily in judging the quality and in determining the degree of bleaching of the oil. For Powe [42], the darker the colour, the poorer the quality. Therefore the pale yellow colour of *Citrullus lanatus* seeds oils shows that the quality of the oil is good and confirms to Encyclopedia of Chemical Technology [42,43].

4. CONCLUSION

The results of the present study showed that ivorian's *Citrullus lanatus* seeds oils contained appreciable physicochemical characteristics which suggest that they could be good sources of edibles oils for cooking. The oils extracted from four varieties of *Citrullus lanatus* revealed that the most abundant fatty acid detected was oleic acid (46.56-59.97). Theses oils contents make them suitable as food supplement. They may also enjoy applications as industrial ingredients in soap production, cosmetics and food complements. Their chemical properties and fatty acids was quite comparable to those of previously reported cucurbitaceae seeds oils, suggesting their potential use as good table and cooking oils. With a high yield of oil and physicochemical characteristics similar to those of the other commercial edible oils, the *Citrullus*

lanatus seeds oils can be considered as a new and valuable source of edible oil. This study also showed potential for *Citrullus lanatus* seeds oils to have relatively high !oxidative stability that would be suitable for food and industrial applications.

COMPETING INTERESTS

Authors have declared that no competing interests exist.

REFERENCES

1. Ng TJ. New Opportunities in *Cucurbitaceae*. In new crops. Janick J, Simon JE. (Eds). Wiley, Newyork. 1993;538-546.

2. Daniel Z, Maria H. Domestication of plants in the old word, third edition (Oxford: University press. 2000;193.

3. Lazos E. Nutritional, fatty acid and oil characteristics of pumpkin and melon seeds. Journal of Food Science. 1986;51:1382-1383.

4. Olaefe O, Adeyemi FO, Adediran GO. Amino acid and mineral composition and functional properties of some oils seeds. J. Agric. Food Chem. 1994;42:878-884.

5. Fokou E, Achu MB, Tchouarguep MF. Preliminary nutritional evaluation of five species of egusi seeds in Cameroun. Afr. J. Food Agric. Nutr. Dev. 2004;4:1-11.

6. Achu MB, Fokou E, Tchiégang C, Fotso M, Tchouanguep FM. Nutritive value of some *Cucurbitaceae* oilseeds from different regions in Cameroun. Afr. J. Biotechnol. 2005;4:1329-1334.

7. Oluba OM, Ogunlowo YR, Ojieh GC, Adebisi KE. Physicochemical properties and fatty acid composition of *Citrullus*

lanatus (Egusi Melon) Seed Oil. Journal of Biological Sciences. 2008;8(4):814-817.

8. AOAC. Official methods of analysis of the association of official analytical chemists. 13th Ed. William Horwitz Ed. 1980;56-132. WASHINGTON D.C. 7.

9. Lozano YF, Dhuique MC, Bannon C, Gaydou EM. Unsaponifiable Matter, total sterol and tocopherol contensts of avocado oil varieties. J. Am. Oil Chem. Soc. 1993;70:561-565.

10. Capannesi P, Mascini M, Parenti A. Electrochemical sensor and biosensor for polyphenols detection in olive oils. Food Chem. 2000;71:553-562.

11. Edidiong A, Eduok UM. Chemical analysis of *Citrullus lanatus seed* oil obtained from Nigeria. Elixir International Journal. 2013;54:12700-12703.

12. Fokou E, Achu MB, Kansci G, Ponka M, Fotso C, Tchiegang FM. Tchouanguep. Chemical properties of some *Cucurbitaceae* oils from Cameroun, Pak. J. Nutr. 2009;8:1325-1334.

13. Umerie SC, Onuagha MC, Nwobi SC. Characterization and utilisation of Adenopus breviflorus Benth (*Cucurbitaceae*) seed oil, Nat. Appl. Sci. J. 2009;10:104-111.

14. Essien EE, Bassey SA, Nimmong-uwem SP. Lagenaria siceraria (Mol.) Standley. Properties of Seeds oils and Variability in Fatty Acids Composition of Ten Cultivars. International Journal of Natural Products Research. 2013;3(4):102-106.

15. Ardabili AG, Farhoosh R, Khodaparast MHH. Chemical composition and physicochemical properties of pumpkin seeds (*Cucurbita pepo* subsp. Pepo Var. Styriaka) grown in Iran. J. Agr. Sci. Tech. 2011;13:1053-1063.

16. Duel HJJr. The lipids: Their Chemistry and Biochemistry. Inter Science Publishers, New York. 1951;1:53-57.

17. Nichols DS, Sanderson K. The nomenclature, structure and properties of food lipids, in:. Sikorski ZE, Kolakowska A (Eds), Chemical and Functional Properties of Food Lipids, CRC Press. 2003;29-59.

18. Meara ML. Fats and other oils, in: Modern methods of plant analysis. Berlin, Gottingen, Heidelberg. 1955;331.

19. Codex Alimentarius. Codex Alimentarius Standards for Fats and oils from Vegetable Sources. Section 2. Codex alimentarius Standards for Named Vegetrable oils. Codex Alimentarius-Stan. 1999;210.

20. Pearson D. The chemical analysis of foods 7th Eden, Churchill Livingstone, London; 1976.

21. Omoti U, Okiy DA. Characteristics and Composition of the Pulp Oil and Cake of the African Pear, Dacryodes edulis (Don, G.) J. Sci. Food and Agric. 1987;38:67-72.

22. Tchiégang C, Kapsue C, Parmenter M. Chemical Composition of oil from Coula edulis (Bail.) nuts. J. Food Lipids. 1998;5:103-111.

23. Kapchie NV, Tchiégang C, Mbofung CM, Kapseu C. Variation dans les caractéristiques physicochimiques des fruits de l'aiélé (*Canarium schweinfruthii* E.) de differentes provenances du Cameroun. 3rd International Workshop on the improvement of safou and other non-conventional oil crops. Yaoundé, Cameroun. Presse Universitaire D'Afrique, ISBN 2-912086-47-7. 2000;249-262.

24. Aremu MO, Olonisakin A, Bako DA, Madu PC. Compositional studies and physicochemical characteristics of cashew nut (Anarcadium occidentale) flour. Pak. J. Nutr. 2006;5:328-333.

25. Ige MN, Ogunsua AO, Okon OL. Functional properties of the protein of some Nigeria oil seeds. Casophor seeds and three varieties of some Nigeria oil seeds. Food Chem. 1984;32:822-825.

26. Badifu GIO. Unsaponifiable Matter in oils from some Species of *Cucurbitaceae*. J. Food Compos. Anal. 1991;4:360-365.

27. Answar F, Answar T, Mahmood Z. Methodical characterization of rice (*Oryza sativa*) Bran oil from Pakistan. Gras. Aceit. 2005;56:125-134.

28. Milovanovic M, Picuric-Jovanovic K. Characteristics and composition of melon seed oil. Journal of Agricultural Sciences. 2005;50:41-47.

29. Savage GP, Mc Neil DL, Dulta PC. Lipid Composition and oxidation stability of oils Hazelnuts (*Corylus avenllana*) grown in New Zealand. Journal of American oil chemists society. 1997;74:755-759.

30. Rethinam P. New Approaches to Coconut, Oils and Fats. African Journal of Food, Agriculture, Nutrition and Development. 2003;19(2):22-25.

31. Pericin D, Krimer V, Trtvic S, Radulovic L. The Distribution of phenolic acids in pumpkin's Hull-Less Seed, Skin Oil Cake Meal, Dehulled Kermel and Hull. Food Chem. 2009;113:450-456.

32. Siger A, Nogala-Kalucka M, Lampart-Szczapa E. The Content and Antioxidant Activity of Phenolic Compounds on Cold-Pressed Plant Oils. J. Foods Lipids. 2008;15:137-149.

33. Boskou D. Sources of Natural Phenolic Antioxidants. Trends Food Sci. Teh. 2006;17:505-512.

34. Moreno JJ, Mitjavila MT. The degree of unsaturation of dietary fatty acids and the development of atherosclerosis (review), J. Nutr. Biochem. 2003;14:182-195.

35. Hu FB, Manson JE, Willett WC. Types of dietary fat and risk of coronary heart disease: A critical review, J. Am. Coll. Nutr. 2001;20:5-19.

36. Idouraine A, Kohlhepp EA, Weber CW. Nutrient constituents from eight lines of naked seed squash (Cucurbita pepo L.), J. Agric. Food Chem. 1996;44:721-724.

37. Zdunczyk Z, Minakowski D, Frejnagel S, Flis M. Comparative study of the chemical composition and nutritional value of pumpkin seed cake, soybean meal and casein, Nahrung. 1999;43:392-395.

38. Fokou E, Achu MB, Kansci G, Ponka R, Fosto M, Tchiegang C, Tchouanguep FM. Chemical properties of some Cucurbitaceae oils from Cameroun, Pak. J. Nutr. 2009;8:1325-1334.

39. Olaofe O, Ogungbenle HN, Akhadelor BE, Idris AO, Omojola OV, Omotehinse OT, Ogunbodede OA. Physicochemical and fatty acids composition of oils from some legume seeds, IJBPAS. 2012;1:355-363.

40. Sadou H, Sabo H, Alma MM, Saadou M, Leger C. Chemical content of the seeds and physico-chemical characteristic of the seed oils from Citrullus colocynthis, Coccinia grandis, Cucumis metuliferus and Cucumis prophetarum of Niger, Bull.Chem. Soc. Ethiop. 2007;21:323-330.

41. FAO. Expert's recomendations of Fats and oils in human nutrition. Fats and oils in human Food and Nutrition Paper. 1994;57-7.

42. Powe WC. Kirk-Othmer Encyclopedia of Chemistry Technology, 3rd ed. New York : Wiley-interscience publishers. 1998;41.

43. Onimawo A, Oyeno F, Orokpo G, Akubor PI. Physicochemical and Nutrient Evaluation of African Bush Mango (Irvingia gabonensis) Seeds and Pulp. Plant Foods for Human Nutrition. 2003;58:1-6.

12

Phytochemical Composition and Antimalarial Activity of Methanol Leaf Extract of *Crateva adansonii* in *Plasmodium berghei* Infected Mice

A. N. Tsado[1*], L. Bashir[2], S. S. Mohammed[2], I. O. Famous[2], A. M. Yahaya[2], M. Shu'aibu[2] and T. Caleb[2]

[1]*Department of Basic and Applied Sciences, Niger State Polythechnic Zungeru, Nigeria.*
[2]*Department of Biochemistry, Federal University of Technology, P.M.B.65, Minna, Niger State, Nigeria.*

Authors' contributions

This work was carried out in collaboration between all authors. All authors read and approved the final manuscript.

Editor(s):
(1) Anonymous.
Reviewers:
(1) Francis W. Hombhanje, Centre for Health Research & Diagnostics, Divine Word University, Papua New Guinea.
(2) Anonymous, Brazil.

ABSTRACT

The need for new compounds active against malaria parasites is made more urgent by the rapid spread of drug-resistance to available antimalarial drugs. The crude methanolic leaf extract of *Crateva adansonii* was investigated for its antimalarial activity against *Plasmodium berghei* (NK65) infected mice. A total of 15 mice were intraperitoneally infected with chloroquine sensitive *P. berghei* strain and divided into 5 equal groups, group 1 served as negative control (untreated), groups 2,3 and 4 were given 200, 400 and 600 mg/kg *Crateva adansonii* methanolic leaves extract respectively while group 5 served as positive control and was given 5 mg\kg chloroquine for five days. The phytochemical constituents of the plant extract were evaluated to elucidate the possibilities of their antimalarial effects. The extract produced a significant dose dependent decrease in the level of parasitaemia when compared to infected untreated group. Also, the extract at dose of 400 mg/kg and 600 mg/kg produced significant increase in body weight and PCV of the infected mice as compare to mice treated with 200 mg/kg of extract and infected untreated group. Phytochemical screening showed that the leaves extract contains alkaloids, anthraquinones,

*Corresponding author: Email: Amos.infonet@gmail.com

tannins, flavonoids, saponins cardiac glycosides and steroids. It is concluded that *Crateva adansonii* could serve as a possible source of antimalarial compounds.

Keywords: Crateva adansonii; antimalarial; phytochemical; parasitarmia; Plasmodium berghei.

1. INTRODUCTION

Malaria is a chronic endemic disease that obstructs social and economic development. It is the leading cause of mortality and morbidity around the word with an estimated 225 million malaria case and 781,000 death being reported globally in 2009 [1]. Most of the reported malaria cases occur in tropical and subtropical regions where the atmospheric condition (temperature and rainfall) are favourable for the development of vectors and parasite [2], the mortality occur mostly in young children and pregnant woman [3]. The attack of malaria during pregnancy stage usually results into severe anemia and impairment of fetal nutrition which contribute to the low birth weight, premature delivery mental retardation and 60% miscarriages [4]. Malaria is caused by protozoan of genus *Plasmodium* transmitted to the vertebrate by female Anopheles mosquitoes in the vertebrate host, the sexual blood forms of the parasite are the life cycle stage that are responsible for the morbidity and mortality of plasmodia infections [5]. Four species of malaria parasite cause disease in human: *Plasmodium falciparum, Plasmodium malariae, Plasmodium vivax* and *Plasmodium ovale* where as 3 species give rise to considerable natural morbidity, only P. *falciparum* result in high mortality [6], as a result of prevalence, virulence and drug resistance. The continuous spread of the *Plasmodium falciparum* resistance to the commonly use anti-malarial drug including the newly introduce Artemisinin Combination Therapy (ACT) has resulted in resurgence in treatment failure [7], and hence the need to intensify research in the area of development of new anti malarial drugs especially from medicinal plant with traditional antimalarial reputation.

Crateva adansonii DC, also known as *Crateva religiosa* or sacred garlic pear, belongs to family Capparaceae. The plant is in high demand, especially its leaves for the treatment of ear infections. The bark is widely used for stomach troubles and held to have tonic properties. In Senegal the roots figure in several treatments for syphyilis, jaundice and fever. Agboke and coworkers claimed the antimicrobial properties of its leaves [8]. Two phytoconstituents had also been isolated and identified as oleanolic acid and 4-epi-hederagenin [9].

In order to scientifically validate the medicinal claimed of *Crateva adansonii* and determine its effectiveness as a potential source of new antimalarial agent, the present study was undertaken to evaluate its phytochemical composition and *in vivo* anti-malarial properties against *Plasmodium berghei* infected mice.

2. MATERIALS AND METHODS

2.1 Plant Collection

Fresh leaves of *Crateva adansonii* were collected from Baddegi, Niger State Nigeria. It was identified and authenticated by a Botanist in the Department of Biological Science, Federal University of Technology Minna, Niger State.

2.2 Experimental Animal

Swiss albino mice weighing between 20-25 g were used in this study. The animals were obtained from the National Institute for Pharmaceutical Research and Development (NIPRD), Idu, Abuja, Nigeria. They were housed in plastic cages with saw dust bed and given standard laboratory diet and water *ad-libitum*. They were then allowed to acclimatize for two weeks to their new environment before the initiation of the experiments

2.3 Parasite

A chloroquine-sensitive strain of *Plasmodium berghei* (NK-65) was obtained from the National Institute for Pharmaceutical Research and Development (NIPRD), Idu, Abuja, Nigeria and maintained by re-infestation via intraperitoneal with infected blood suspension (0.2 ml) containing about 1×10^7 suspension of *P. berghei* parasitized red blood cells.

2.4 Sample Preparation and Extraction Procedure

The collected fresh leaves of *Crateva adansonii* was washed with clean-water and air dried. The dried sample was grounded using a grinder mill. Extraction of plant material was performed by soxhlet extraction using methanol. The resulting methanol extract was concentrated in a water bath and stored in a refrigerator until required.

2.5 Determination of Yield of Extract

The percentage yield of methanol leaf extract of *Crateva adansonii* was determined by weighing the coarse sample before extraction and the methanol leaf extract of *Crateva adansonii* after concentration and then calculated using the formula.

Percentage yield (%) =

Weight (g) of the concentrated extract X 100
Weight (g) of the *Crateva adansonii* leaf

2.6 Phytochemical Analysis

Methanol leaves extract of *Crateva adansonii* was characterized for phytochemical composition including alkaloids, anthraquinones, tannins, flavonoids, saponins, cardiac glycosides, steroids and phlobatannins according to the methods Harborne, [10], and Sofowora [11].

2.7 *In vivo* Antimalarial Study

2.7.1 Curative test

Evaluation of the curative potential of the crude extract in Peter's test was carried out according to the method described by Ryley and Peters [12]. On Day 0, standard inocula of 1×10^7 infected erythrocytes were inoculated in mice intraperitoneally. Seventy-two hours later, mice were randomly divided into their respective groups and dosed accordingly once daily for five days. The extract was dissolve in normal saline.

Group I Mice were given 200 mg/kg b.w of methanol leaves extract of *C. Adansonii.*
Group II Mice were given 400 mg/kg b.w of methanol leaves extract of *C. Adansonii.*
Group III Mice were given 600 mg/kg b.w of methanol leaves extract of *C. Adansonii.*
Group IV Mice received 5 mg chloroquine /kg body weight.
Group V Mice were given normal saline/kg body weight.

2.7.2 Daily parasitaemia count

On each day a drop of blood were collected from the tail of each rat, smeared unto a microscopic slide to make thin films, stained with 10% Giemsa stain and examined microscopically to monitor the parasitaemia level.

2.7.3 Determination of packed cell volume (PCV)

The capillary tubes were filled with blood to about 1cm or two-third (2/3) of its length and the vacant end of each of the capillary tubes was sealed by plastic seal or sealer to protect the blood level from spilling. The tubes were placed in haematocrit centrifuge with seal side towards the periphery and then centrifuge for 5-6 minutes. The percentage of packed cell volume or haematocrit was read directly from haematocrit reader [13].

2.8 Statistical Analysis

Data were analyzed using statistical package for social science (SPSS) version 16 and presented as means±SEM. Comparisons between different groups was done using Analysis of Variance (ANOVA) and Duncan's Multiple Range Test (DMRT). Values of P<0.05 were considered as statistically significant as described by Mahajan, [14].

3. RESULTS

3.1 Extract Yield

The percentage yield of leaves extract of *Crateva adansonii* is shown in Table 1. The yield of methanol leaves extract of *Crateva adansonii* was 25.09%.

3.2 Phytochemicals

Table 2 shows the result of qualitative phytochemical composition of methanol leaf extract of *Crateva adansonii*. The results revealed the present of alkaloids, anthraquinones, tannins, flavonoids, saponins cardiac glycosides and steroids, however, phlobatannins were absent.

Table 1. The percentage (%) yield of methanolic leaves extract of *Crateva adansonii*

Crateva adansonii	Weight (g)
Leaf powder	50.00
Methanol extract	12.547
Extract yield (%)	25.09

Table 2. Qualitative phytochemical composition of methanol leaf extract of *Crateva adansonii*

Phytochemicals	Inference
1. Alkaloids	++
2. Cardiac glycosides	++
3. Anthraquinones	++
4. Steroids	++
5. Tannins	+
6. Saponins	+++
7. Flavonoids	++
8. Reducing sugars	++
9. Phlobatannins	-

Key: (-) absent, (+) slightly present, (++) moderately present, (+++) highly present

3.3 Antimalarial Study

3.3.1 Parasitaemia count

The average daily parasitaemia level of the *Plasmodium berghei* infected mice treated with methanol leaves extract of *Crateva adansonii* are shown in Fig. 1 and Table 3. The average daily parasitaemia of infected mice treated with methanol leaves extract of *Crateva adansonii* at doses of 400 and 600mg/kg were significantly (P<0.05) reduced (37.71 and 40.41% inhibition)

when compare with the negative control over the period of the experiment. The average daily parasitaemia of infected mice treated with chloroquine was significantly (P<0.05) reduced (54.51% inhibition) when compared with extract treated groups. However no significant (p>0.05) difference in the level of parasitaemia count of infected mice treated with 200mg/kg leaf extract of *Crateva adansonii* as compared with the control group.

3.3.2 Body weight changes

Effect of methanol leaves extract of *Crateva adansonii* on body weight of *Plasmodium berghei* infected mice is shown in Fig. 2 the body weight of the infected untreated mice and infected treated with 200 mg/kg of *Crateva adansonii* show significant decrease in body weight after 5 days of treatment. However the infected mice treated with 400 mg/kg, 600 mg/kg of *Crateva adansonii* as well as those treated with 5 mg/kg chloroquine show significant increase in body weight after 5 days of treatment.

3.3.3 Packed cell volume

Effect of methanol leaves extract of *Crateva adansonii* on PCV of *Plasmodium berghei* infected mice are shown in Fig. 3 the PCV of *P. berghei* infected untreated mice and infected treated with 200 mg/kg of *Crateva adansonii* show significant decrease in PCV after 5 days of treatment. However the infected mice treated with 400 mg/kg, 600 mg/kg of *Crateva adansonii* as well as those treated with 5 mg/kg chloroquine show significant increase in PCV after 5 days of treatment.

Table 3. Average parasitaemia count of *Plasmodium berghei* infected mice treated with methanol leaves extract of *Crateva adansonii*

Groups	Day 3	Day 4	day 5	Day 6	Day 7	% parasite reduction
Chloroquine	44.40±3.45[a]	32.60±3.11[a]	22.10±2.23[a]	15.12±2.45[a]	3.50±0.47[a]	54.51
200 mg/kg C. adansonii	44.30±2.14[a]	44.76±4.56[b]	51.97±3.45[c]	56.82±4.34[c]	58.98±5.66[c]	-
400 mg/kg C. adansonii	43.10±4.31[a]	37.90±4.44[a]	33.10±3.45[b]	29.70±2.34[b]	17.4±1.19[b]	37.71
600 mg/kg C. adansonii	44.10±4.03[a]	35.30±3.60[a]	31.90±2.50[b]	28.20±2.11[b]	14.70±1.34[b]	40.41
control	46.20±3.45[a]	47.00±3.67[b]	50.9±6.35[c]	55.68±4.01[c]	59.01±5.67[c]	-

Values are M±SEM of triplicate determination; Values along the same column with different superscripts are significantly different (p < 0.05)

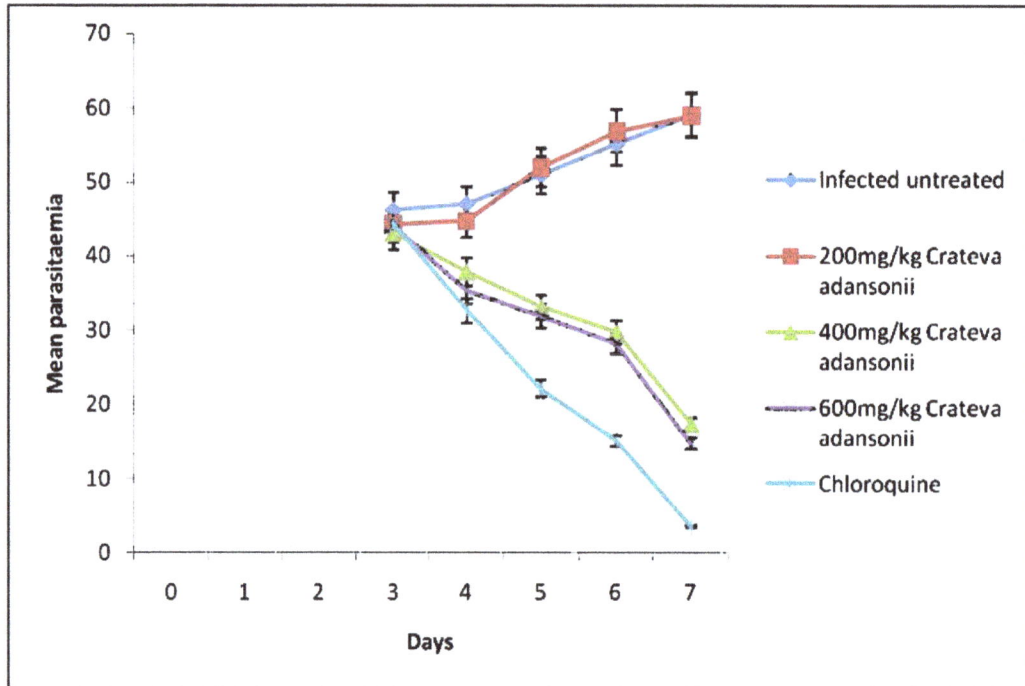

Fig. 1. *In vivo* antiplasmodial activity of methanol leaves extract of *Crateva adansonii* against *Plasmodium berghei infected mice:* Each point is a Mean±SEM of triplicate determination

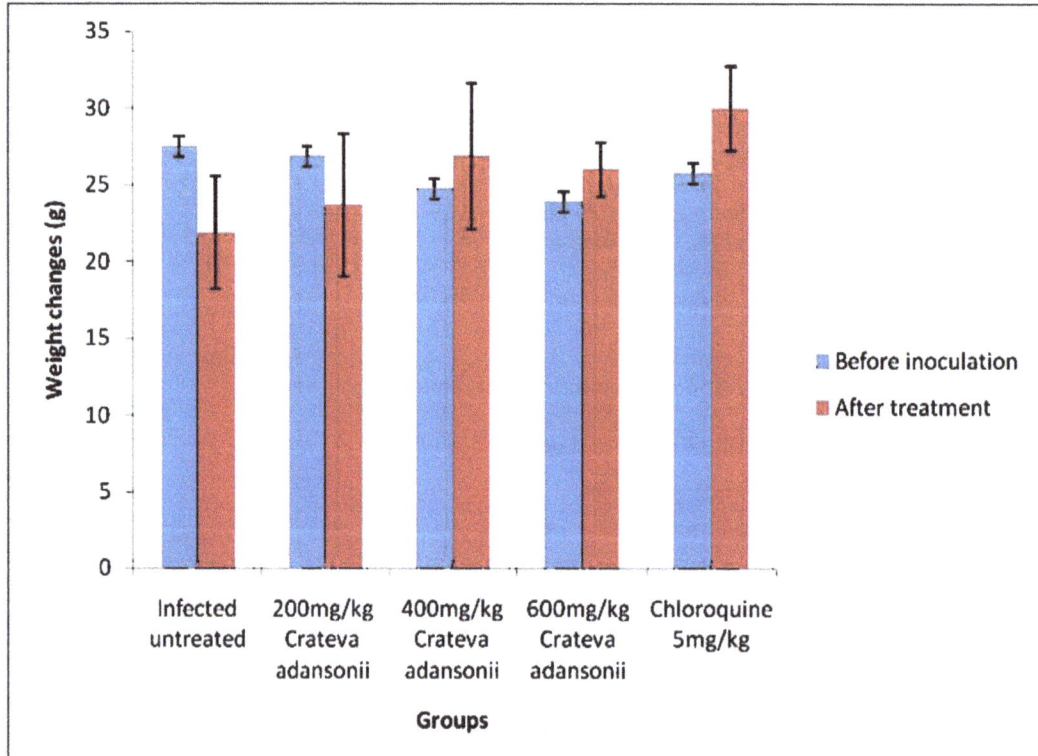

Fig. 2. Effect of methanol leaf extract of *Crateva adansonii* on body weight of *P. berghei* infected mice. Values are M±SEM of triplicate determination

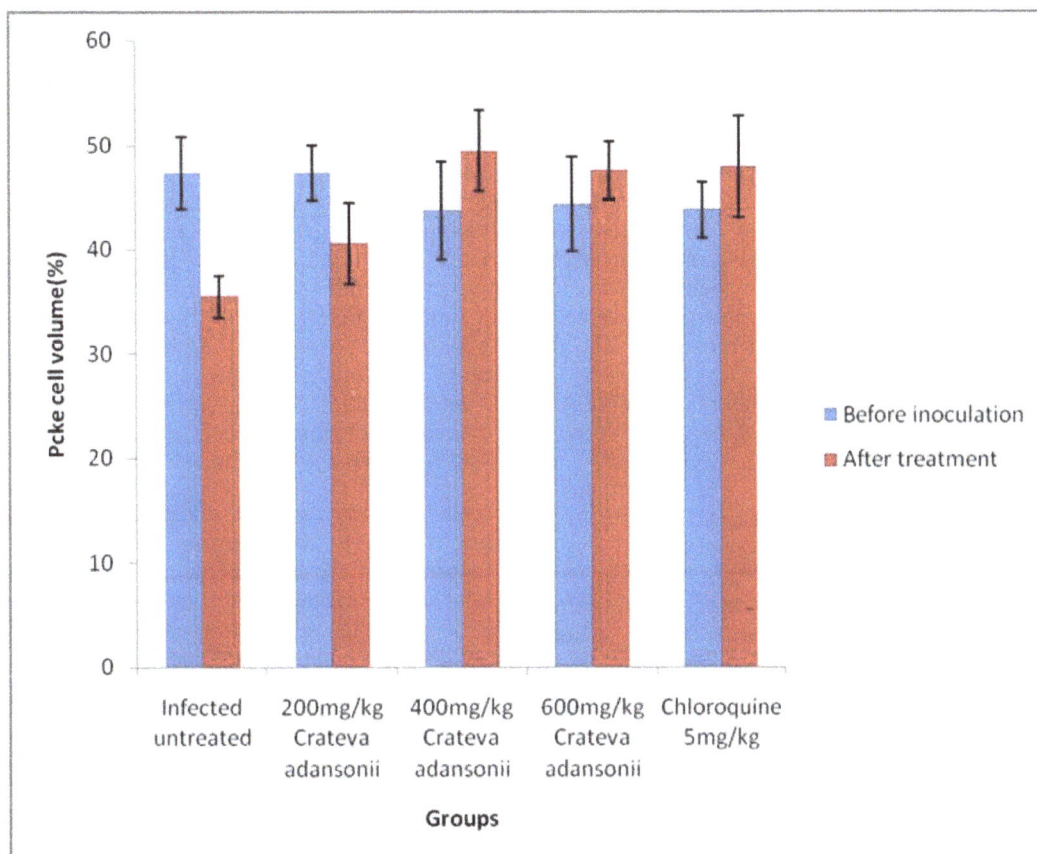

Fig. 3. Effect of methanol leaves extract of *Crateva adansonii* on PCV of *P. berghei* infected mice. Values are M±SEM of triplicate determination

4. DISCUSSION

Plants used in treatment of diseases are said to contain active compounds called phytochemicals some of which are responsible for their characteristic adours, pugencies and colour while others give to a particular plant its virtues as food, medicinal or poisonous [15]. There is considerable interest by phytochemists to identify the therapeutic agent contained in a medicinal plant in order to establish the basis for their uses in traditional medical practice.

This study revealed the presence of various medicinal important phytochemicals including alkaloids, anthraquinones, tannins, flavonoids, saponins cardiac glycosides and steroids in methanolic leaves extract of *Crateva adansonii*.

Flavonoids are compounds with a widespread occurrence in the plant kingdom which have also been detected in *Artemisia* species. They are reported to have exhibited significant *in vitro* antimalarial activity against *P. falciparum* [16]. Their presences in *Crateva adansonii* extract could justify the antimalarial activities exhibited by the plant extract.

Saponins are used as adjuvants in the production of vaccins [17]. Steroids are used in the stimulation of bone marrow and growth. They stimulate lean body mass and also play vital roles in the prevention of bone loss in elderly men [18]. Alkaloids have been used as CNS stimulant, topical anesthetic in ophthalmology, powerful painkillers, and antipyretic action among other use [19]. The cardiac glycoside has been used for over two centuries as stimulant in cases of cardiac failure and diseases [20]. The presence of tannins in the leaves extract of *Crateva adansonii* suggests the ability of these plants to play major roles as antifungal, antidiarrheal, antioxidant and antihemorrhoidal agent [21]. Tannins also have astringent properties, plants containing tannins have been reported to be used for healing of wounds, varicose ulcers, hemorrhoids, frostbile and burn in herbal medicine [22].

The presence of all these phytochemcials in the leaves extract of *Crateva adansonii* is an indication that this plant, if properly screened, could yield a drug of pharmacological significance. However the absence of phlobatannins agree with early studies which also found that not all phytochemicals are present in all plants and those present differ with the solvent used in the extraction process [23].

The *in vivo* antimalarial effect of *Crateva adansonii* against *Plasmodium berghei* infected mice was evaluated. The extract showed a significant dose dependent and progressive reduction in parasitaemia with time, this is a very promising feature in the potentiality of *Crateva adansonii* as an antimalarial drug. However, the antilamalarial effect demonstrated by *Crateva adansonii* leaves extract was lower when compared to chloroquine. Chloroquine has been used as the standard antimalarial drug because of its established activities on *P. berghei* [24]. The *P. berghei*, a rodent malarial parasite although not able to infect man and other primate has been used in antimalarial assays because of its sensitivity to chloroquine [25]. Also the insignificant difference in the level of parasite count of infected mice treated with 200 mg/kg of the extract when compared with the control group reflect the inactivity of the extract which could be attributed to low concentration of bioactive agents at that doses.

Anemia, body weight loss and body temperature reduction are the general features of malaria-infected mice [26]. So, an ideal antimalarial agents obtained from plants are expected to prevent body weight loss in infected mice due to the rise in parasitemia. In the present study, extract of *C. adansoni* significantly prevented weight loss associated with increase in parasitemia level.

Blood parameters including packed cell volume (PCV) is used to assess anemia, erythrocytosis, hemodilution and hemoconcentration due to disease condition [13]. One of the major reasons for the development of anemia is oxidative stress [27], the immune system of the body is activated by malarial infection thereby causing the release of free radicals. In addition to this the malaria parasite also stimulates certain cells to produce reactive oxygen species (R.O.S) their by resulting in haemoglobin degradation [28]. Thus,

the significant decrease level of PCV and body weight of the untreated mice is an indication of anemic condition caused by the malarial infection. The significant increase in level of PCV and body weight in mice treated with *Crateva adansonii* at 400 and 600 mg/kg when compare with the control group is an indication of ameliorating potentials of the plant extract on the anaemia induced by the malarial infection. Though the rodent malaria model, *P. berghei* is not similar to that of human Plasmodium parasite the antiparasitic activities demonstrated by the extract especially at doses of 600mg/kg against *P. berghei* mice in this study could be an indication that the extract could possibly be effective against human malarial parasite [29].

The antimalarial activity demonstrated by this plant might be attributed to the presence of alkaloids or flavonoids which have been identified presently in this work; or even a combined action of more than one metabolite. However, the active compound(s) known to give this observed activity need to be identified.

5. CONCLUSION

The study suggests that *C. adansonii* leaf contains important phytoconstituents that could be implicated in the observed antimalarial effect of the plant. The study also represents an important preliminary demonstration of the potential of this antimalarial herbal medicine to treat human malaria. However, a bioguided fractionation of the methanolic extract of *C. adansonii* leaves aiming to isolating active compound(s) is reccomended.

CONSENT

It is not applicable.

ETHICAL CLEARANCE

Ethical Clearance was given by Federal University of Technology, Minna/Nigerian Ethical Review Board (CUERB) in accordance with International standard on care and use of experimental animals.

ACKNOWLEDGEMENTS

Thanks are due to the National Institute for Pharmaceutical Research and Development, Abuja, Nigeria for providing Chloroquine sensitive *P. berghei* (NK65) strain that was used in this study

COMPETING INTERESTS

Authors have declared that no competing interests exist.

REFERENCES

1. USAID. Saving lives: A global leader in fighting malaria. USAID: (2011). From the American People. Infectious diseases. Retrieved:http://ec.europa.eu/research/health/infectious-diseases/poverty diseases/malaria_en.html

2. Greenwood BM, Fidock DA, Kyle DE, Kappe SH, Alonso PL, Collins FH, Duffy PE. Malaria: progress, perils, and prospects for eradication. Journal of Clinical Investigation. 2008;118:1266-1276.

3. Ogunlana OO, Ogunlana OE, Ademowo, OG. Comparative in vitro assessment of the antiplasmodial activity of quinine - zinc complex and quinine sulphate. Scientific Research and Essay. 2009;4(3):180-184.

4. WHO. The global malaria situation: current tools for prevention and control, Geneva. 2002;1-5.

5. Ahmed KB. Antibody responses in *Plasmodium fakiparum* malaria and their relation to protection against the disease a thesis from department of immunology, the wenner-green institute stockholm university, stockholm, Sweden; 2004

6. WHO. World malaria situation in 1997 part 1. Weekly epidemiological record WHO Geneva. 1997;72:269-274

7. Oseni LA, Akwetey GM. An in-vivo evaluation of antiplasmodial activity of aqueous and ethanolic leaf extracts of *Azadiractha indica* in *Plasmodium berghei* infected BALB/c mice. International Journal of Pharmaceutical Sciences and Research. 2012;3(5):1406-1410.

8. Agboke A, Attama A, Momoh A. Evaluation of the antimicrobial properties of crude extract of *Cryptolepis sanguinolenta* and *Crateva adansonii* leaves and their interactions. Journal of Applied Pharmaceutical Science. 2011;1(10):85-89

9. Cantrell MA, Berhow BS, Phillips SM, Duval D, Weisleder SF, Vaughn CL. Bioactive crude plant seed extracts from the NCAUR oilseed repository. Phytomedicine. 2003;10:325-333.

10. Harborne JB. Methods of plant analysis. In: Phytochemical Methods. Chapman and Hall, London; 1973.

11. Sofowora A. Medicinal plants and Traditional Medicine in Africa. John Wiley and Sons Ltd, New Tork. 1982;80-96.

12. Ryley JF. Peters W. The antimalarial activity of some quinoline esters. Ann Trop Med Parasitology. 1995;84:209-222.

13. Dacie JV, Lewis SM. Practical Haematology. 9th Edition Churchill Livingstone; 2000.

14. Mahajan BK. Significance of differences in means. In: Methods in Biostatistics for Medical and Research Workers, 6th edition. New Delhi: JAYPEE Brothers Medical Publishers. 1997;130-155.

15. Evans WC. Trease and Evans pharmalognosy (15th edition) W.B Saunders company LTD. London. 2002;191-393.

16. Chanphen R, Thebtaranonth Y, Wanauppathamkul S, Yuthavong Y. Antimalarial principles from *Artemisia indica*. Journal of Natural Products. 1998;61:1146-1147

17. Asl M, Hossein H. Review of Pharmacological effects of Glycyrrhiza specie and its bioactive compounds. Phytochemistry. 2008;66:24-709

18. Depiccoli BI, Giada F, Benettin A, Sarton F, Piccoli E. Anabolic steroid use in body builders an echocardiograhic study of left ventricle morphology and function. International Journal of Sport Medicine. 1991;12:12-408.

19. Njoku PC, Akumefula MI. Phytochemical and nutrient evaluation of Spondias mombin Leaves. Pak. J. Nutr. 2007;6(6): 613-615.

20. Trease E, Evans, WC. Pharmacognosy. Billiare tindall London. 13th edition; 1987; 61-62

21. Asquith TN, Butter LG. Interaction of condensed tannins with selected proteins. Phytochemistry. 1986;25(7):1591-1593.

22. Igboko DO. Phytochemical studies on *Garcinia Kola* Heckel. M. Sc Thesis. University of Nigeria, Nsukka. 1983;202.

23. Tijjani IM, Bello I, Aliyu A, olunnshe T, Logun, Z. Phytochemical and antibactenanl study of root extract *Cochlospermum tinctoricm*. American Research Journal of Medicinal Plant. 2007;3:16-22.

24. Ajaiyeoba E, Falade M, Ogbole O, Okpako L, Akinboye D. *In vivo* Antimalarial and Cytotoxic Properties of Annona Senegalensis Extract. African Journal of Traditional Medicine. 2006;3(1):137-141.

25. Fidock DA, Rosenthal PJ, Croft SL, Brun R, Nwaka S. Antimalarial drug discovery: Efficacy models for compound screening Supplementary documents. Trends in Parasitology. 2004;(15):19–29.

26. Langhorne J, Quin SJ, Sanni LA. Mouse models of blood-stage malaria infections: Immune responses and cytokines involved in protection and pathology. In Malaria Immunology. 2nd edition. Edited by Perlmann P, Troye-Blomberg M. Stockholm: Karger publisher. 2002;204–228.

27. Kremsner PG, Greve B, Lell B, Luckner D, Schmidt D. Malarial Anaemia in African Children Associated with High Oxygen-Radical Production. Lancet. 2000;355:40-41.

28. Loria P, Miller S, Foley M, Tilley L. Inhibition of peroxidative degradation of heme as the basis of action of chloroquine and other quinoline antimalarials. Biochem. J. 1999;339:363-370.

29. Audualem G. Evaluation of antimalarial activity of seeds of *Dodonae angustifolia* against P. Berghei in mice a thesis submited to school of graduate studies of the Addis Ababa University; 2000.

13

Protective Effects of CYP2E1 Inhibitors on Metabolic Syndrome-induced Liver Injury in Guinea Pigs

Volodymyr V. Rushchak[1*], Ganna M. Shayakhmetova[2], Anatoliy V. Matvienko[2] and Mykola O. Chashchyn[1]

[1]*Department of Molecular Oncogenetics, Institute of Molecular Biology and Genetics, NAS of Ukraine, Zabolotnogo Str. 150, Kyiv, Ukraine.*
[2]*Department of General Toxicology, SI "Institute of Pharmacology and Toxicology, NAMS of Ukraine", Eugene Pottier 14, Kyiv, Ukraine.*

Authors' contributions

This work was carried out in collaboration between all authors. Author VVR managed the analyses of the study, performed the statistical analysis and wrote the first draft of the manuscript. Author GMS managed the analyses of the study, managed the literature searches and wrote part of the manuscript. Author AVM managed the histopathology studies. Author MOC designed the study and wrote the protocol. All authors read and approved the final manuscript.

Editor(s):
(1) Ge Qiang, University of Texas Southwestern Medical Center at Dallas, Dallas, Texas, USA.
Reviewers:
(1) A. Papazafiropoulou, Department of Internal Medicine and Diabetes Center, Tzaneio General Hospital of Piraeus, Greece.
(2) Jaspinder Kaur, Contributory Health Scheme (ECHS) Polyclinic, Sultanpur Lodhi, Kapurthala District 144626, India.
(3) Ds Sheriff, Faculty of Medicine, Benghazi University, Benghazi, Libya.

ABSTRACT

The present work reports the effects of CYP2E1-inhibitors (quercetin, 4- methylpyrazole and disulfiram) on the indices characterizing state of the liver in guinea pigs with metabolic syndrome (MS) induced by protamine sulfate repeated administrations.
The investigation of quercetin, 4-methylpyrazole and disulfiram effects on hepatic cytochrome P450 2E1 (CYP2E1) protein and activity changes was conducted. Simultaneously, the content of reactive oxygen species (ROS) and markers of liver damage were determined in experimental animals' blood. The link between increased hepatic expression of CYP2E1, prominent ROS generation and liver damage in animals with MS has been discovered. It has been demonstrated that CYP2E1 protein content and activity in guinea pigs with MS rose almost 3 times compared to intact animals. These events were accompanied by increase in ROS generation and metabolism

Corresponding author: Email: v.v.rushchak@gmail.com

and liver disturbances symptoms: increase in serum glucose and cholesterol contents (2 and 2.6 times respectively), alanine aminotransferase (2.8 times), aspartate aminotransferase (6.4 times), and alkaline phosphatase (1.8 times) elevations. Our investigation suggests that administration of quercetin, 4-methylpyrazole, and disulfiram in guinea pigs with MS caused decrease in this isoenzyme protein expression (2.5, 1.7 and 2.4 respectively) as well as its enzymatic activity in liver (6.6, 1.1, and 1.7 respectively). The content of blood ROS was partially restored or normalized by all three CYP2E1 inhibitors. In turn, suppression of CYP2E1 activity and ROS generation led to decrease in hepatic MS manifestation. It is apparent from the present observation that quercetin has the highest efficiency among the investigated substances. Further studies on various quercetin doses and administration regimens could provide relevant information for the development of MS-related nonalcoholic fatty liver disease treatment.

Keywords: CYP2E1; metabolic syndrome; diabetes; oxidative stress; protamine sulphate.

1. INTRODUCTION

The prevalence of metabolic syndrome (MS) is growing around the world at an alarming rate. MS is defined by a constellation of interconnected physiological, biochemical, clinical and metabolic factors that directly increases the risk of cardiovascular disease, type 2 diabetes mellitus and all cause mortality. Insulin resistance, visceral adiposity, atherogenic dyslipidemia, endothelial dysfunction, genetic susceptibility, elevated blood pressure, hypercoagulable state, and chronic stress are the several factors which constitute the syndrome [1].

Nonalcoholic fatty liver disease (NAFLD) is the hepatic expression of MS, which comprises a spectrum of clinical and histological events ranging from simple and benign fatty liver to steatohepatitis, which is characterized by the abnormal activation of pathways leading to an aggressive inflammatory condition [2,3]. NAFLD is usually clinically silent, and its impact has most likely been underestimated. Symptoms of NAFLD, if present, are minimal and non-specific, such as fatigue and right upper quadrant discomfort. The disease usually comes to medical attention incidentally when aminotransferases levels are found to be elevated or a radiographic study reveals that the liver is fatty [4]. This pathological state may progress to more severe damage known as cirrhosis, which endangers the anatomy and function of liver tissue. In addition, a small group of patients with end-stage liver disease may develop hepatocellular carcinoma and finally death [3].

It is generally assumed that NAFLD is commonly associated with insulin resistance, which is a major risk factor for the development of type 2 diabetes and a central feature of the MS [5]. Hyperinsulinemia, caused by an increased insulin secretion by the pancreatic β-cells and decreased insulin degradation by the liver, is a compensatory phenomenon to insulin resistance. Hyperinsulinemia leads to an increase in fat mass, lipogenesis and it is associated with increased concentrations of free fatty acids [6]. Most of the adverse effects induced by fatty acids accumulation are likely to be mediated by lipid intermediates, notably diacylglycerols and ceramides. Lipid intermediates can induce insulin resistance by activating different kinases such as mammalian target of rapamycin (mTOR), inhibitor of κB kinase (IKK), Jun N-terminal kinase (JNK) and novel protein kinase C (nPKC) that are known to exert negative feedback on proximal insulin signaling [5].

Indeed, insulin resistance and oxidative stress are major pathogenic mechanisms leading to chronic liver diseases in subjects with MS. The oxidative stress includes reactive oxygen species (ROS) production and lipid peroxidation eventually causing NAFLD [7]. The potential sources for the ROS in the liver include hepatic cytochrome P450 2E1 (CYP2E1), mitochondria and iron overload [7]. Insulin resistance and increased cytochrome CYP2E1 expression are both associated with and mechanistically implicated in the development of liver pathology. Although currently viewed as distinct factors, insulin resistance and CYP2E1 expression may be interrelated through the ability of CYP2E1-induced oxidant stress to impair hepatic insulin signaling [8]. On the other hand, CYP2E1 is normally suppressed by insulin but is invariably increased in the livers of patients with NAFLD [9]. Once liver disease is established, NAFLD-induced impairment of insulin signaling may then further promote the diabetic state. Findings of reduced suppression of hepatic glucose production by insulin in NAFLD patients as compared with controls support this concept. There is suggestion that one mechanism of this effect can be mediated by the effects of CYP2E1 over-expression [8].

On the assumption of the above mentioned, the regulation of CYP2E1 expression level could be effective method to alleviate the extent of pathological processes in the liver accompanying MS. Recent findings, showing that severe liver injury associated with elevated oxidative stress was blunted by inhibitors of CYP2E1, also stimulated our interest in this problem [10]. Based on these facts and considering that MS has emerged as one of the major health care issues, the present work reports the effects of CYP2E1-inhibitors (quercetin, 4-methylpyrazole (4-MP), and disulfiram) on the indices characterizing state of the liver in guinea pigs with MS.

2. MATERIALS AND METHODS

2.1 Animals and Experimental Design

Male guinea pigs (n = 25) with initial mean body weight 370g (5 months old) were used in the study. They were kept under a controlled temperature (from 22°C to 24°C), relative humidity of 40% to 70%, lighting (12 h light-dark cycle) and on a standard pellet feed diet ("Phoenix" Ltd., Ukraine) and water ad libitum.

2.2 Chemicals

All reagents used in investigations were obtained from Sigma Aldrich and Serva.

The following CYP2E1 inhibitors were applied: protamine sulfate (Protamine sulfate salt from salmon) was from Serva; quercetin (3,5,7,3',4'-pentahydroxyflavone), 4-MP (4-Methyl-1H-pyrazole) and disulfiram (1,1',1",1'''-[disulfanediylbis (carbonothioylnitrilo)] tetraethane) were from Sigma Aldrich.

2.3 Experimental Design

The guinea pigs were kept for acclimatization during 10 days, and then they were randomized into 5 groups. Each group contained 5 animals:

1. Control, n=5: intact animals.
2. MS, n=5: guinea pigs were injected intramuscularly with protamine sulfate solution in dose 15 mg/kg b.w., twice per day, daily during 5 weeks. After this animals

were kept in usual conditions during 4 weeks till euthanasia [11].
3. MS with quercetin administration, n=5: Animals with MS were given quercetin (intramuscularly, 20 mg/kg b.w., daily, during 2 weeks). Quercetin administration was started in 2 weeks after the last protamine sulfate injection. Euthanasia of animals was performed in overnight after the last quercetin administration.
4. MS with 4-MP administration, n=5. Animals with MS were given 4-MP (intramuscularly, 20 mg/kg b.w., daily, during 2 weeks). Administration was started in 2 weeks after the last protamine sulfate injection, and euthanasia was performed in overnight after the last 4-MP administration.
5. MS with disulfiram administration, n=5. Animals with MS were given disulfiram (per os, 25 mg/kg b.w., daily, during 2 weeks). Administration was started in 2 weeks after the last protamine sulfate injection, and euthanasia was performed in overnight after the last disulfiram administration.

2.4 Blood and Tissue Collection

All overnight fasted animals were sacrificed in above mentioned terms under a mild ether anesthesia by decapitation. Euthanasia always was performed at the same time to avoid the daily changes in the activity of enzymes.

Blood for biochemical tests and flow cytofluorometry was withdrawn from the femoral vein prior to euthanasia. Blood with heparin was used for flow cytofluorometry analysis. Non-heparinized blood was centrifuged (1300 RCF, RT, and 10 minutes) in order to separate serum.

Liver was removed and processed for morphological studies.

Also samples of liver (100 mg) were collected, quickly frozen in liquid nitrogen, and stored at -80°C before examination.

2.5 Western Blot Analysis

The relative level of CYP2E1 protein in the liver was determined by Western blot analysis.

The extraction of total proteins was carried out according to a standard protocol for membrane proteins (Abcam, UK).

Proteins from the liver of each animal (50 μg per line) were separated using 12% polyacrylamide gel with 0.1% sodium dodecyl sulfate (SDS). The semi-dry electro transfer of proteins to the nitrocellulose membranes was held at 200 mA for 40 minutes. Western blot analysis was held in the following way: nitrocellulose membranes (Biorad, USA) were pre-incubated in 2% nonfat milk (Sigma, USA), and then treated with polyclonal anti-CYP2E1 antibodies (obtained in Molecular Oncogenetics Department, Institute of Molecular Biology and Genetics NASU) at a 1:400 ratio v/v for 1 hour. After washing membranes were incubated with secondary anti-Rabbit IgG-HRP antibodies (Sigma, USA) at a 1:5000 ratio v/v during the same time. The β-actin (used as internal control) was identified using anti-β-actin antibodies (Sigma-Aldrich, USA). The treatment of membranes with secondary antibodies was followed by chemiluminescence detection according to manufacturers' instructions (Pierce). Membranes were exposed to autoradiography film (Agfa, Belgium) for 0.5 to 1 minutes. Digital images of immunoblots were analyzed using densitometric scanning analysis program Scion image 3.53.346.0 (http://www.scioncorp.com/).

The level of CYP2E1 protein was calculated as the ratio of protein values to β-actin on the same line, and presented as reference units.

2.6 Evaluation of CYP2E1 Activity in Liver Microsomes

Microsomes were obtained according to the protocol described by Jeong and Yun [12].

A spectrophotometric method for determination of CYP2E1 activity by monitoring of the *p*-nitrocatechol formation from *p*-nitrophenol (PNP) by isolated liver microsomes was used. This method is applicable to enzymatic studies for determination of P450-catalyzed *p*-nitrophenol hydroxylation activity [13]. The enzymatic product, *p*-nitrocatechol, is assayed at 546 nm after acidification of the reaction mixture with trichloroacetic acid followed by neutralization using 10 M NaOH.

2.7 Cytofluorometric Analysis of ROS

Studies were performed as it was described by Bhagwat et al. [14] using a Coulter Epics XL Flow cytometer (Beckman Coulter, US) with argon laser. The excitation wavelength was 488 nm. Detection channels were FL1 (515-535 nm) and FL3 (620-630 nm). The dyes used in this study (25 µmol/L 2,7-dichlorodihydrofluorescein

diacetate (2,7-DCFH-DA) and 10 µg/ml propidiun iodide) were from Molecular Probes (Leiden, the Netherlands).

The results were presented in reference units of dichlorodihydrofluorescein (oxidized product 2,7-DCFH-DA) fluorescence.

2.8 Clinical Chemistry

Glucose, and cholesterol contents, alanine aminotransferase (AlAT), aspartate aminotransferase (AsAT) and alkaline phosphatase (ALP) activities in blood serum were determined with fully automatic biochemistry analyzer (Prestige 24i, Tokyo Boeki, Japan).

2.9 Histopathology

Liver was fixed in 10% buffered neutral formalin, dehydrated in ethanol solutions and embedded in paraffin. Histologic sections (4 µm) were stained by McManus's method for glycogen detection [15].

Microscopic studies were carried out with microscope Cytophan (Leica Microsystems, Wetzlar GmbH).

2.10 Statistical Analysis

The obtained data were calculated by one-way analysis of variance (ANOVA) and expressed as mean ± standard error of mean (SEM). Data were compared using Tukey test. Differences were considered to be statistically significant at $P < 0.05$.

3. RESULTS

3.1 Evaluation of Liver CYP2E1 Protein Level and Enzymatic Activity

Western Blot analysis was performed to evaluate the effect of MS on CYP2E1 protein expression in the liver. We demonstrated that the CYP2E1 protein content in guinea pigs with MS increased more than 3 times in comparison with intact animals (Fig. 1A and 1C).

PNP hydroxylase activity as a selective enzyme marker for CYP2E1 was determined. The obtained results (Fig. 1B) suggested the significant increase in CYP2E1 enzymatic activity following MS development. It was also almost 3 times greater than in control (Fig. 1B).

As it is shown in Fig. 1A and 1B, CYP2E1 inhibitors in varying degrees reduced its expression and activity in the liver. The most effective inhibition was observed following quercetin administration, the least effective – following 4-MP treatment. It is interesting that 4-MP more effectively inhibited CYP2E1 protein expression, than its activity, whereas quercetin acted as the most powerful inhibitor of the enzyme activity.

3.2 Cytofluorometric Analysis of ROS

In order to explore the influence of MS-mediated CYP2E1 induction in liver on oxidative stress development, the cytofluorometric analysis of ROS level in experimental animals' leukocytes was performed.

It was shown that the development of MS in the experimental animals accompanied by a considerable increase in the number of blood ROS, which content was higher than in intact group almost 4 times (Fig. 2).

Data represented in Fig. 2 suggest that reducing of activity and CYP2E1 protein level by inhibitors led to decrease (2 times) in the intensity of ROS generation.

3.3 Clinical Biochemistry and Histopathology

Clinical biochemistry parameters and liver histology were studied in order to evaluate severity of MS signs in animals of different experimental groups.

In animals groups with MS we have revealed symptoms typical for this pathology, such as hyperglycemia and hyperlipidemia. Serum glucose and cholesterol contents were increased 2 and 2,6 times respectively, as compared with control (Table 1).

At the same time significant increases in liver damage marker enzymes activities were fixed (AlAT and AsAT – 2.8 and 6.4 times, ALP – 1.8 times).

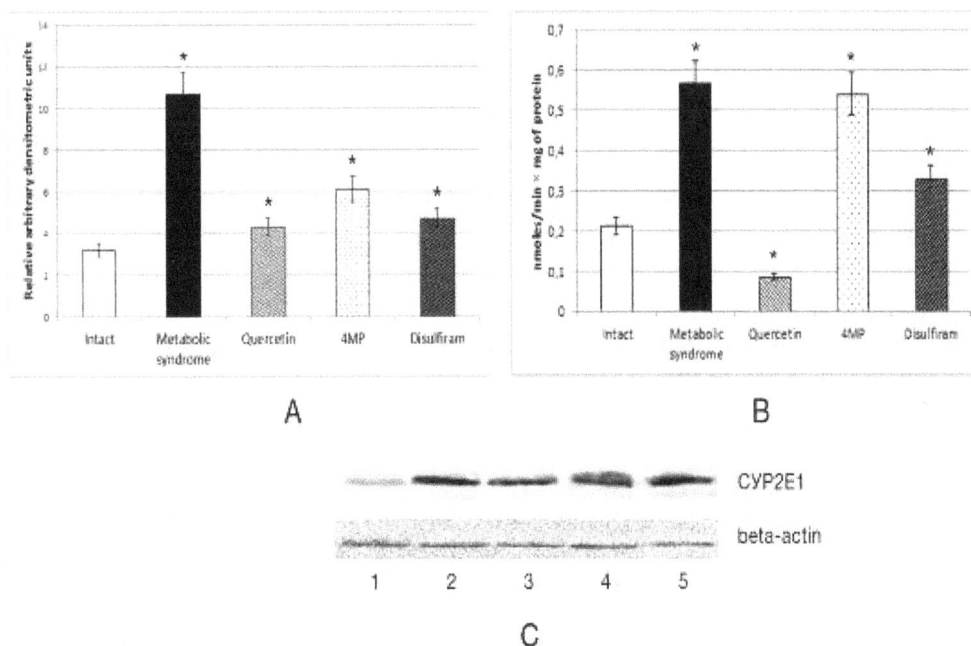

A

B

C

Fig. 1. Average rate of CYP2E1 protein expression, (M±S.E.M., n=5) (panel A) and CYP2E1 enzymatic (PNP hydroxylase) activity, M ± S.E.M., n=5 (panel B) in liver of guinea pigs with MS and CYP2E1 inhibitors administration. Representative Western Blot of CYP2E1 and reference-gene β-actin proteins are shown in panel C. Line 1 – intact animals, Line 2 – animals with MS, Line 3 – MS treated by quercetin; Line 4 – MS treated by 4MP; Line 5 – MS treated by disulfiram
– P<0.05 in comparison with control

These results are in concordance with our data of liver histopathology. The main pathogenetic mechanism of MS development is carbohydrate metabolism disturbance. Normally liver cells contain a large amount of glycogen (Fig. 3A, intensive pink staining). Reducing of the livers slice color intensity in MS group (Fig. 3B) evidences the common for MS and DM 2 type [16,17] decrease of glycogen content due to glycogenolysis activation.

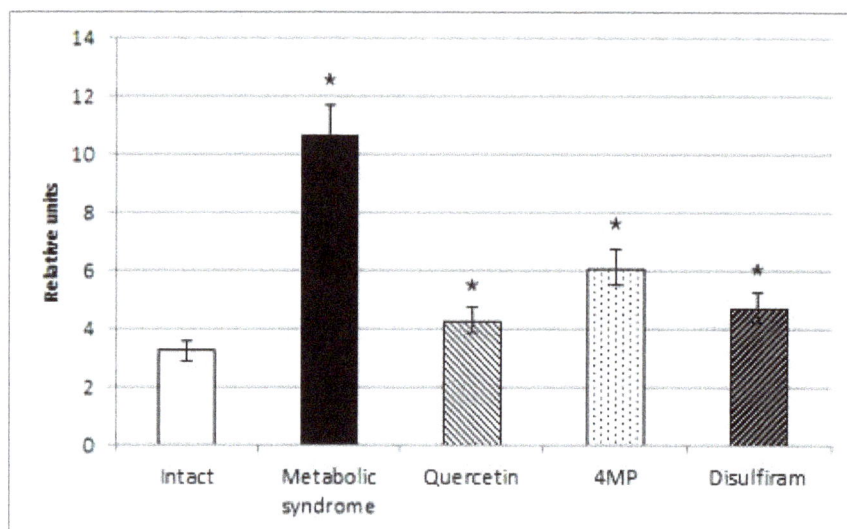

Fig. 2. Average content of ROS (M ± S.E.M., n=5) in leukocytes of guinea pigs with MS and CYP2E1 inhibitors administration

– P<0.05 in comparison with control

Table 1. Clinical biochemistry parameters in guinea pigs with MS and CYP2E1 inhibitors administration

Groups of animals	Parameters				
	Glucose, mol/L	Cholesterol, mol/L	AlAT, IU/L	AsAT, IU/L	ALP, IU/L
Control	5.2±0.2	0.58± 0.03	37.5±2.6	39.5±2.7	68.75±1.2
MS	10.2±0.3*	1.5±0.09*	107.0±6.2*	255.0±15.1*	126.0±4.1*
MS + quercetin	8.65±0.3*	0.71±0.06*	59.5±3.5*	80.0±4.4*	81.0±2.8*
MS + 4-methyl-pyrazole	9.1±0.3*	0.72±0.07*	86.0±4.7*	176.0±12.7*	113.0±3.1*
MS + disulfiram	8.1±0.2*	0.79±0.09*	85.0±4.3*	201.0±11.2*	108.5±4.4*

– P<0.05 in comparison with control

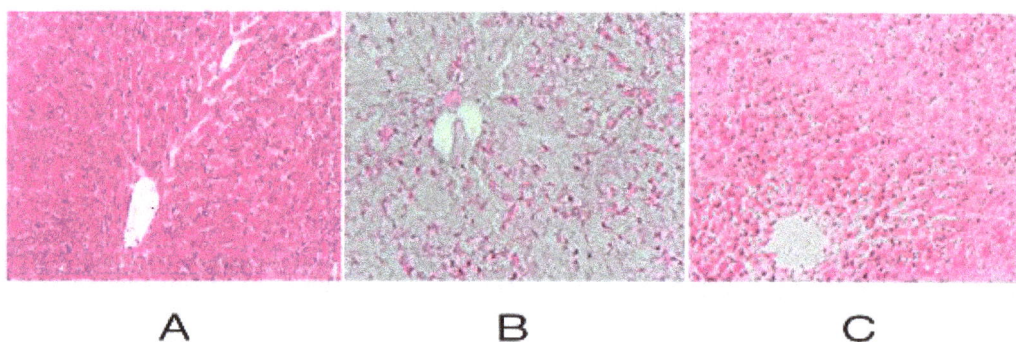

Fig. 3. Histochemical determination of glycogen content in liver cells (PAS-reaction by McManus). x1000, x400. A - control; B - MS; C - MS + quercetin

To identify links between CYP2E1-dependent processes at MS and liver state, we compared the serum biochemical parameters depending on the severity of CYP2E1 inhibition by disulfiram, 4-MP or quercetin. It was shown that CYP2E1 inhibitors alleviated hepatic manifestation of MS to varying degrees (Table 1). According to our findings all three compounds prevented hyperglycemia development. Serum glucose content in treated guinea pigs was almost 2 times lower than in MS group (Table 1). Furthermore, we have clearly indicated the most pronounced effect of the quercetin administration exhibiting marked reduction in AlAT, AsAT and ALP activities as compared with non-treated animals (Table 1). The partial restoration of glycogen content in liver cells was observed only following quercetin administration (Fig. 3.C). Disulfiram and 4-MP had no effects at this parameter (data are not shown).

4. DISCUSSION

It is known that cytochrome P450 system is a powerful source of ROS generation [18,19]. The contribution of isoform CYP2E1 is particularly noticeable, due to its high oxidase activity and ability to produce ROS even in the absence of the substrate [20]. Among them a special role belongs to superoxide anion (O_2^-), a byproduct of the CYP2E1-mediated metabolism [21], which can serve as part of the second hit to advance the severity of NAFLD. It is seems well documented that CYP2E1 protein expression and activity increases in obesity, fatty liver, and NASH in both humans and rodents, and this increase appears to correlate well with the severity of NAFLD [22]. On the other hand, the NAFLD has been described as the hepatic manifestation of MS. Main risk factors associated with MS are abdominal obesity, insulin resistance, diabetes and dyslipidemia, but it is of interest, that NAFLD can be described in non-obese and non-diabetic patients [23].

In our experiments the induction of CYP2E1 protein expression in MS-group guinea pigs' liver was accompanied by increasing in its enzymatic activity in this organ. Such results are in coincidences with other authors' data indicating the increase in CYP2E1 mRNA and protein levels in both obese and diabetic humans [24,25].

It should be noted that not only CYP2E1-dependent O_2^- production [21,26] contributes to liver injury. It has been reported recently that CYP2E1-mediated oxidative stress causes M1 macrophages polarization bias, which includes a significant increase in interleukin-1β (IL-1β) and IL-12 in experimental models of NASH, whereas CYP2E1-null mice or diallyl sulfide administration prevented it [27]. Our results on increase of ROS generation in organisms of guinea pigs with MS are in good accordance with above mentioned reports and suggest about oxidative stress occurrence, which could be one of critical factors in the development of NAFLD. Indeed, in serum of experimental animals we registered indicators of metabolic disturbances and liver injury, such as increases in glucose and total cholesterol contents, elevations of AlAT, AsAT, and ALP activities. Significant decrease in glycogen content in hepatocytes of MS-group also is clear evidence of MS-related liver pathology. As it has been established by Samuel et al. [28], at NAFLD, hepatic fat accumulation alone is insufficient to increase endogenous glucose production, but that it does cause hepatic insulin resistance. This can be attributed in part to decreased insulin-stimulated tyrosine phosphorylation of insulin receptor substrate 1 (IRS-1) and IRS-2, which in turn blocks the ability of insulin to activate glycogen synthase and diminishes the ability of the liver to store glucose as glycogen.

Our attention was attracted to the report on protection from a high-fat diet-induced insulin resistance in $Cyp2e1^{-/-}$ mice [29,30]. Conversely, mice knocked in for the human $CYP2E1$ transgene had higher fasting insulin level, greater hepatic fat accumulation, high level of oxidant stress, and liver injury [31]. The ability of insulin to decrease CYP2E1 expression has been proved, as well as the fact that insulin resistance lead to increase of CYP2E1 expression and activity [32]. In this case, the mechanism of CYP2E1 induction can be realized via the high concentration of ketone bodies produced from persistent mitochondrial fatty acid oxidation. Ketone bodies stabilize CYP2E1 and prevent its degradation. The increase in CYP2E1 and enhanced insulin resistance seem to promote each other by creating a positive feedback loop that may eventually make steatosis progress to steatohepatitis as oxidant stress increases [33]. Recently Leung and Nieto [33] have suggested that inhibiting of CYP2E1 may disrupt this feedback loop and reduce insulin resistance and liver injury. Moreover, recent work has shown that CYP2E1 activity correlates with ethanol-induced liver injury and lipid peroxidation [34]. The use of CYP2E1 inhibitors, such as chlormethiazole and polyenylphosphatidylcholine, demonstrate partial but effective protection in ethanol-induced liver injury [35,36].

In this regard it was of our interest to investigate the different CYP2E1 inhibitors as tools for correction of MS liver's manifestations.

We have used three inhibitors of CYP2E1 (disulfiram, 4-MP and quercetin) to determine link between this isoenzyme expression and level of liver injury. Above mentioned substances have rather different inhibitory mechanisms toward CYP2E1.

There is report that disulfiram is not directly responsible for inactivation of CYP2E1, but its reduced form, diethyldithiocarbamate, reacts with the enzyme leading to its inactivation [37]. Authors have postulated that this metabolite could inactivate CYP2E1 by forming a disulfide bond with one of eight cysteines that are present in the apoprotein [37].

The CYP2E1 inhibitor 4-MP competed with PNP for the CYP2E1 catalytic site as shown through catalytic, binding, and docking studies [38]. Direct evidence for the high affinity interaction between the catalytic site and 4-MP has been shown through the generation of the type II binding spectra between CYP2E1 and the inhibitor. When bind to the catalytic site, 4-MP mediates van der Waals contacts with the same residues as observed for PNP. An additional interaction with Leu-210 and formation of the Fe–N bond likely plays a role in the higher selectivity for 4-MP over PNP hydroxylation [38].

As it has been shown by other authors, inhibition of CYP2E1 by 4-MP significantly decreases oxidative stress [39]. The property of 4-MP to inhibit CYP2E1 and alcohol dehydrogenase is widely used to prevent poisoning with ethanol, methanol and ethylene glycol [39- 41].

Quercetin, one of the most common flavonoids presents in various vegetables, fruits, herbs and red wine, and possesses broad bioactivity which is based on or implicated in its prominent antioxidative properties [42]. Growing experimental data have demonstrated that quercetin exhibits ability to suppress CYP2E1 activity [43-46]. *In vitro* experiments have shown that CYP2E1 suppression is concomitant with heme oxygenase-1 (HO-1) induction by quercetin and protects hepatocytes from ethanol-induced

oxidative damage [47,48]. Further research of the same authors has demonstrated that HO-1 induction and CYP2E1 down-regulation is accompanied with decreased heme pool [44]. Depleted heme pool and CO release may contribute to the protective mechanism of quercetin by limiting the protein synthesis and directly inactivating of heme-containing CYP2E1 [44].

Our investigation suggests that administration of all three substances in guinea pigs with MS causes decrease of hepatic CYP2E1 protein expression, as well as its enzymatic activity. Our results, nevertheless, point to significant variability in their inhibitory effects toward PNP hydroxylation activity, while protein expression is suppressed approximately at the same level. The most pronounced effect on CYP2E1 activity was revealed in case of quercetin administration. Conversely, 4-MP administration practically was not effective. Such data are in agreement with results of other authors demonstrating relatively small impact of 4-MP on CYP2E1 activity [49,50]. These authors have shown also the 4-MP ability to decrease the content of ROS in hepatocytes cultures treated with ethanol [49,50].

The regulation of CYP2E1 expression and activity is of importance since CYP2E1 plays a central role in MS-related liver injury. CYP2E1 is triggered by both exogenous and endogenous substrates, and it induces disturbances in hepatocytes by generating ROS and lipid peroxidation reactions. The decline of endogenous antioxidants in such conditions may further enhance CYP2E1-induced lipid peroxidation, oxidant stress and cellular toxicity [8]. As expected, in our experiments down-regulation of CYP2E1 activity in liver of animals with MS led to reduce of oxidative stress in their organism. The number of blood ROS were partially restored or normalised by quercetin, 4-MP, and disulfiram treatment.

It is widely accepted that under pathologies accompanying by CYP2E1 induction, its inhibition and decrease in ROS generation are beneficial for maintenance of structure and function of liver state [51]. Indeed, our results on clinical chemistry parameters and liver histology confirm this theory. Decrease in CYP2E1 protein content and activity, as well the level of ROS led to reduction of liver cells damage. At least, following quercetin administration the liver state almost returned to normal. It is possible that such remarkable hepatoprotective activity of quercetin

Protective Effects of CYP2E1 Inhibitors on Metabolic Syndrome-induced Liver Injury in...

127

can be realized not only due to its inhibitory effect on CYP2E1, but also because of its well-known antioxidant properties [42]. This assumption is confirmed by the fact that following 4-MP and disulfiram administration we have observed only tendency to AlAT, AsAT and ALP decrease and no effects on liver glycogen content.

5. CONCLUSION

Collectively our results suggest that liver damage in animals with MS may be reversible. The link between increased hepatic expression of CYP2E1, prominent ROS generation and liver damage in animals with MS has become clear. In turn, suppression of CYP2E1 activity using specific inhibitors leads to decrease in hepatic MS manifestation. However, further studies are warranted to understand the most effective ways to regulate hepatotoxic and hepatoprotective pathways by CYP2E1 inhibitors. From the present observation is apparent that quercetin has the highest efficiency among the investigated substances but its most effective doses remain unclear. Further studies on various quercetin doses and administration regimens could provide relevant information for the development of MS-related NAFLD treatments.

ETHICAL APPROVAL

The "Principles of laboratory animal care" (NIH publication No. 85-23, revised 1985) were followed, as well as specific national laws where applicable. All experiments have been examined and approved by the bioethics committees of the Institute of Molecular Biology and Genetics, NAS of Ukraine and the SI "Institute of Pharmacology & Toxicology", NAMS of Ukraine.

ACKNOWLEDGEMENTS

The authors gratefully acknowledge the financial support of this Study by the National Academy of Sciences of Ukraine.

The team is grateful to the technical staff of Molecular Oncogenetics Department of Institute of Molecular Biology & Genetics, NAS of Ukraine, and General Toxicology Department of SI "Institute of Pharmacology & Toxicology NAMS of Ukraine" for their necessary assistance during the study.

COMPETING INTERESTS

Authors have declared that no competing interests exist.

REFERENCES

1. Kaur J. Comprehensive review on metabolic syndrome. Cardiology Research and Practice. 2014;2014:1-21.

2. Levene AP, Goldin RD. The epidemiology, pathogenesis and histopathology of fatty liver disease. Histopathology. 2012;61(2): 141-52.

3. Medina-Santillán R, López-Velázquez JA. Hepatic manifestations of metabolic syndrome. Diabetes Metab Res Rev. 2013;7. Avalaible:http://onlinelibrary.wiley.com/doi/10.1002/dmrr.2410/abstract

4. Kim CH, Younossi ZM. Nonalcoholic fatty liver disease: A manifestation of the metabolic syndrome. Cleveland Clinic Journal of Medicine. 2008;75(10):721-28.

5. Asrih M, Jornayvaz FR. Inflammation as a potential link between nonalcoholic fatty liver disease and insulin resistance. J Endocrinol. 2013;218(3):25-36.

6. Bugianesi E, McCullough AJ, Marchesini G. Insulin resistance: a metabolic pathway to chronic liver disease. Hepatology. 2005; 42(5):987-00.

7. Paradies G, Paradies V, Rugierro FM, Petrosillo G. Oxidative stress, cardiolipin and mitochondrial dysfunction in nonalcoholic fatty liver disease. World J Gastroenterol. 2014;20(39):14205-18.

8. Schattenberg JM, Wang Y, Singh R, Rigoli RM, Czaja MJ. Hepatocyte CYP2E1 over expression and Steatohepatitis lead to impaired hepatic insulin signaling. The Journal of Biological Chemistry. 2005; 280(11):9887-94.

9. Chitturi S, Farrell GC. Etiopathogenesis of nonalcoholic steatohepatitis. Semin Liver Dis. 2001;21(1):27-41.

10. Cederbaum A. CYP2E1 potentiates toxicity in obesity and after chronic ethanol treatment. Drug Metabol Drug Interact. 2012;27(3):125-44.

11. Rushchak VV, Kovalenko VM, Voronina AK, Kitam VO, Maksymchuk OV, Chashchyn MO. Optimization of animal model for investigation of the type 2 diabetes pathogenesis. Fiziol. Journal. 2012;58(6):29-35. Ukrainian.

12. Jeong HG, Yun C-H. Induction of rat hepatic cytochrome P450 enzymes by myristicin. Biochem Biophys Res Commun. 1995;217(3):966-71.

13. Reinke LA, Moyer MJ. p-Nitrophenol hydroxylation. A microsomal oxidation

which is highly inducible by ethanol. Drug Metab Dispos. 1985;13(5):548–52.

14. Bhagwat SV, Vijayasarathy C, Raza H, Mullick J, Avadhani NG. Preferential effects of nicotine and 4-(N-methyl-N-nitrosamine)-1-(3-pyridyl)-1-butanone on mitochondrial glutathione S-transferase A4-4 induction and increased oxidative stress in the rat brain. Biochem Pharmacol. 1998;56(7):831–39.

15. Sarkisov D, Perov J, editors. Microscopically technique. Moscow: Medicine; 1996.

16. Magnusson I, Rothman DL, Katz LD, Shulman RG, Shulman GI. Increased rate of gluconeogenesis in type II diabetes mellitus. A 13C nuclear magnetic resonance study. J Clin Invest. 1992;90(4): 1323–27.

17. Shulman GI, Rothman DL, Jue T, Stein P, DeFronzo RA, Shulman RG. Quantitation of muscle glycogen synthesis in normal subjects and subjects with non-insulin-dependent diabetes by 13C nuclear magnetic resonance spectroscopy. N Engl J Med. 1990;322(4):223–28.

18. Yang CS, Yoo JS, Ishizaki H, Hong JY. Cytochrome P450IIE1: roles in nitrosamine metabolism and mechanism of regulation. Drug Metab Rev. 1990;22(2-3):147–59.

19. Guengerich FP, Kim DH, Iwasaki M. Role of human cytochrome P-450IIE1 in the oxidation of many low molecular weight cancer suspects. Chem Res Toxicol. 1991; 4(2):168–79.

20. Surbrook Jr. SE, Olson, MJ. Dominant role of cytochrome P450 2E1 in human hepatic microsomal oxidation of the CFC-substitute 1,1,1,2-tetrafluoroethane. Drug Metab Dispos. 1992;20(4):518–24.

21. Abdelmegeed MA. Critical role of cytochrome P450 2E1 (CYP2E1) in the development of high fat-induced non-alcoholic steatohepatitis. Journal of Hepatology. 2012;57(4):860–66.

22. Aubert J. Increased expression of cytochrome P450 2E1 in nonalcoholic fatty liver disease: Mechanisms and pathophysiological role. Clinics and Research in Hepatology and Gastroenterology. 2001;35(10):630–37.

23. De Araújo Souza MR, de Fátima Formiga de Melo Diniz M, de Medeiros-Filho JEM, de Araújo MST. Metabolic syndrome and risk factors for non-alcoholic fatty liver disease. Arq Gastroenterol. 2012;49(1):89-96.

24. Lieber CS. CYP2E1: from ASH to NASH. Hepatol Res. 2004;28(1):1-11.

25. Leclercq I, Horsmans Y, Desager JP, Pauwels S, Geubel AP. Dietary restriction of energy and sugar results in a reduction in human cytochrome P450 2E1 activity. Br J Nutr. 1999;82(4):257-62.

26. Lieber CS. Cytochrome P450 2E1: its physiological and pathological role. Physiol Rev. 1997;77(2):517-44.

27. Seth RK, Das S, Pourhoseini S, Dattaroy D, Igwe S, Ray JB, et al. M1 Polarization bias and subsequent nonalcoholic steatohepatitis progression is attenuated by nitric oxide donor DETA NONOate via inhibition of CYP2E1-induced oxidative stress in obese mice. J Pharmacol. Exp. Ther. 2015;352:77-89.

28. Samuel VT, Liu Zh-X, Qu X, Elder BD, Bilz S, Befroy D, et al. Mechanism of hepatic insulin resistance in non-alcoholic fatty liver disease. The Journal of Biological Chemistry. 2004;279(31):32345–53.

29. Abdelmegeed MA, Banerjee A, Yoo SH, Jang S, Gonzalez FJ, Song BJ. Critical role of cytochrome P450 2E1 (CYP2E1) in the development of high fat-induced non-alcoholic steatohepatitis. J Hepatol. 2012; 57(4):860–66.

30. Zong H, Armoni M, Harel C, Karnieli E, Pessin JE. Cytochrome P-450 CYP2E1 knockout mice are protected against high-fat diet-induced obesity and insulin resistance. Am J Physiol Endocrinol Metab. 2012;302(5):532–39.

31. Kathirvel E, Morgan K, French SW, Morgan TR. Overexpression of liver-specific cytochrome P4502E1 impairs hepatic insulin signaling in a transgenic mouse model of nonalcoholic fatty liver disease. Eur J Gastroenterol Hepatol. 2009;21(9):973–83.

32. Woodcroft KJ, Hafner MS, Novak RF. Insulin signaling in the transcriptional and posttranscriptional regulation of CYP2E1 expression. Hepatology. 2002;35(2):263–73.

33. Leung T-M, Nieto N. CYP2E1 and oxidant stress in alcoholic and non-alcoholic fatty liver disease. Journal of Hepatology. 2013; 58(2):395–398.

34. Bell LN, Temm CJ, Saxena R, Vuppalanchi R, Schauer P, Rabinovitz M, et al. Bariatric surgery-induced weight loss reduces hepatic lipid peroxidation levels and affects hepatic cytochrome P-450 protein content. Ann. Surg. 2010;251(6):1041–48.

35. Aleynik MK, Leo MA, Aleynik SI, Lieber CS. Polyenylphosphatidylcholine opposes the increase of cytochrome P-4502E1 by ethanol and corrects its iron-induced decrease. Alcohol Clin Exp Res. 1999; 23(1):96–100.

36. Gouillon Z, Lucas D, Li J, Hagbjork AL, French BA, Fu P, et al. Inhibition of ethanol-induced liver disease in the intragastric feeding rat model by chlormethiazole. Proc Soc Exp Biol Med. 2000;224(4):302–08.

37. Pratt-Hyatt M, Lin HL, Hollenberg PF. Mechanism-based inactivation of human CYP2E1 by diethyldithocarbamate. Drug Metab Dispos. 2010;38(12):2286-92.

38. Collom SL, Laddusaw RM, Burch AM, Kuzmic P, Perry MD Jr, Miller GP. CYP2E1 substrate inhibition. Mechanistic interpretation through an effector site for monocyclic compounds. J Biol Chem. 2008;283(6):3487–96

39. Sommerfeld K, Zielińska-Psuja B, Przystanowicz J, Kowalówka-Zawieja J, Orłowski J. Effect of 4-methylpyrazole on antioxidant enzyme status and lipid peroxidation in the liver of rats after exposure to ethylene glycol and ethyl alcohol. Pharmacological Reports. 2012; 64(6):1547–53.

40. Zakharova S, Navratilb T, Pelclova D. Fomepizole in the treatment of acute methanol poisonings: experience from the czech mass methanol outbreak 2012-2013. Biomed Papers. 2014;158(4):641-49.

41. Chen T-H, Cuo C-H, Huang C-T, Wang W-L. Use of Fomepizole in pediatric methanol exposure: the first case report in Taiwan and a literature review. Pediatrics and neonatology; 2013.
Available:http://dx.doi.org/10.1016/j.pedneo.2013.08.009 (In press).

42. Boots AW, Haenen GR, Bast A. Health effects of quercetin: from antioxidant to nutraceutical. Eur J Pharmacol. 2008; 585(2-3):325-37.

43. Kitam VO, Chashchyn MO. Computer modelling of human cytochrome P450 2E1 complex. Ukr Biochem Journal. 2010; 82(2):94-103. Ukrainian.

44. Tang Y, Tian H, Shi Y, Gao C, Xing M, Yang W, et al. Quercetin suppressed CYP2E1-dependent ethanol hepatotoxicity via depleting heme pool and releasing CO. Phytomedicine. 2013;20(8–9):699–704.

45. Surapaneni KM, Priya VV, Mallika J. Pioglitazone, quercetin and hydroxy citric acid effect on cytochrome P450 2E1 (CYP2E1) enzyme levels in experimentally induced nonalcoholic steatohepatitis (NASH). European Review for Medical and Pharmacological Sciences. 2014;18(18): 2736-41.

46. Oliva J, Bardag-Gorce F, Tillman B, French BA, French SW. The effects of dietary supplements on reducing ethanol induced oxidative stress. The FASEB Journal. 2011;25:794-2.

47. Yao P, Nussler A, Liu L, Hao L, Song F, Schirmeier A, Nussler N. Quercetin protects human hepatocytes from ethanol-derived oxidative stress by inducing heme oxygenase-1 via the MAPK/Nrf2 pathways. J Hepatol. 2007;47(2):253–61.

48. Yao P, Hao L, Nussler N, Lehmann A, Song F, Zhao J, et al. The protective role of HO-1 and its generated products (CO, bilirubin, and Fe) in ethanol-induced human hepatocyte damage. Am J Physiol Gastrointest Liver Physiol. 2009;296(6): 1318–23.

49. Gong P, Cederbaum A. Nrf2 is increased by CYP2E1 in rodent liver and HepG2 cells and protects against oxidative stress caused by CYP2E1. Hepatology. 2006; 43(1):144–53.

50. Yang SP, Medling T, Raner GM. Cytochrome P450 expression and activities in the rat, rabbit and bovine tongue. Comp Biochem Physiol C Toxicol Pharmacol. 2003;136(4):297-308.

51. Jaeschke H, Gores GJ, Cederbaum AI, Hinson JA, Pessayre D, Lemasters JJ. Mechanisms of hepatotoxicity. Toxicol Sci. 2002;65(2):166-76.

14

Histopathological Alterations Induced by *Naja naja* Crude Venom on Renal, Pulmonary and Intestinal Tissues of Mice Model

M. A. Al-Mamun[1], M. A. Hakim[2], M. K. Zaman[1], K. M. F. Hoque[1], Z. Ferdousi[1] and M. Abu Reza[1*]

[1]*Protein Science Lab, Department of Genetic Engineering and Biotechnology, University of Rajshahi, Rajshahi- 6205, Bangladesh.*
[2]*Functional Proteomics and Natural Medicines, Kunming Institute of Zoology, the Chinese Academy of Sciences No. 32 Jiaochang Donglu, Kunming, Yunnan, 650223, China.*

Authors' contributions

This work was carried out in collaboration between all authors. Authors MAAM, MAH and AR designed the study, performed the statistical analysis, wrote the protocol, and wrote the first draft of the manuscript. Authors MKZ and ZF managed the analyses of the study. Author KMFH managed the literature searches. All authors read and approved the final manuscript.

Article Information

DOI: 10.9734/BBJ/2015/16188
Editor(s):
(1) Giuliana Napolitano, Department of Biology (DB), University of Naples Federico II, Naples, Italy.
Reviewers:
(1) Rodrigo Crespo Mosca, Institute of Energetic and Nuclear Research - Biotechnology Center Research (University of São Paulo), Brazil.
(2) Anonymous, Colombia.

ABSTRACT

Aim: Snake bite causes a significant number of mortality and morbidity throughout the world. So, the current study was carried out to estimate the extinct of damage caused by intraperitoneal introduction of cobra venom on kidney, lung and intestinal tissues of mice model using histological technique.
Place and Duration of Study: The entire study including the treatment along with preparing histological slide was conducted in protein science laboratory, Department of Genetic Engineering and Biotechnology, University of Rajshahi, Bangladesh between December 2013 to July 2014.
Methods: Twenty five mature female albino mice were divided mainly into two groups as control and envenomated group. Lyophilized *Naja naja* venom was dissolved in 0.9% NaCl solution and injected intraperitoneally into the mice of the envenomated group at dosages equivalent to LD_{50}

---

*Corresponding author: Email: rezaru@gmail.com

(0.25 mg/kg). Whereas the animals from control group were not received any venomous component. Both groups of animal were sacrificed for histological study and visualized under light microscope.

Results: Injection of cobra venom induced a range of histological changes in all envenomated mice comparing with their control. Results from the histopathological examination showed mainly inflammatory cellular infiltration, vacuolation in renal tubules, shrinking of glomeruli, raising space between the walls of Bowman's capsule in renal tissue and alveolar haemorrhage, inflammatory cellular infiltration and edema in pulmonary tissue. No significant histopathological alterations in intestinal tissue were observed without infiltration and mild hemorrhage.

Conclusion: The findings from the current study revealed that, cobra venom at lethal dose causes multiple organ failure in experimental animal which could be considered among the factors that lead to death. By observing the site and the mode of action on tissue level, these findings may help to allay the severity of damage by discovering novel anti venom drug.

Keywords: Cobra; histology; intestine; kidney; lung; Mus musculus and venom.

ABBREVIATIONS

SRCC (Snake Rescue and conservation centre); NaCl (Sodium chloride); LD$_{50}$ (Lethal dose); PLA2 (Phospholipase A2).

1. INTRODUCTION

Snake bite is a familiar term throughout the world especially in the rural areas of Bangladesh and causes a significant clinical impact on human health. Snakebite frequency has been estimated at about 500,000 and mortality between 30000–40000 annually throughout the world [1]. However, the degrees of envenomation related injury relies mainly on the nature of toxic polypeptide in the venom. Venom is a highly modified salivary secretion mainly used to initiate the breakdown of food stuff into soluble compounds for proper digestion [2]. Biochemically snake venom is a complex mixture of several protein riches substances including toxins, enzymes, growth factors, activators and inhibitors with broad spectrum biological activities, stored in venom glands at the back of the head [3-5]. The venomous components which are mainly responsible for severe clinical injury of the victims are pro-coagulant enzymes, anticoagulant enzymes, cytolytic or necrotic toxins, hemolytic and myolytic phospholipases A2, pre- and post-synaptic neurotoxins, and haemorrhagins [6]. A number of these toxic proteins singly or jointly interact with components of important physiological system of envenomated individual and producing diverse clinical effects [7].

Cobra (*Naja naja*) is the most venomous snake of Elapidae family, responsible for a significant number of deaths yearly in our country. Elapidae comprises approximately 325 species of venomous snake, including cobras, mambas, sea snakes and coral snakes widely distributed in the tropical and subtropical regions of the world, especially in southeastern Asia [8]. The symptoms of cobra envenomation mainly are local pain, severe swelling, blistering, necrosis and other non-specific effects [6]. Cobra envenoming cause's multiple-organ failure resulting death in case of severe envenomation [9]. The neurotoxin of cobra venom exerts action by blocking the transmission of nerve signals to muscles leading to the subsequent stopping of nerve transmission to the heart and lungs and death occurs rapidly due to complete failure of respiratory function. Histological, histochemical and biochemical fluctuations triggered by the venom of African black-necked spitting cobra (*Naja nigricollis*) and Egyptian cobra (*Naja haje*) have already been evaluated on rodent animal [10-12]. Whereas toxic effects of *Naja naja* on tissue level have not ever been recorded in the Bangladesh. Since *Naja naja* is the most venomous snake in this region and their venom compounds are slightly different from the other places. So, there was a paramount need to evaluate the mode of action of cobra venom on tissue level of the most important organs of human body using rodent animal as model. Hence, the current study was aimed to detect the intensity of damage resulting intraperitoneal injection of cobra venom on rodent animal.

2. MATERIALS AND METHODS

2.1 Experimental Animals

A total of twenty five female albino mice (26-30 gm) were collected from the Department of

Pharmacy of Jahangirnagar University, Dhaka, Bangladesh. The mice were kept in metal cages (5 per cage) and allowed to feed ordinary poultry food and tap water and kept at the animal house of Biochemistry Department, Faculty of Science, Rajshahi University. All mice were handled in accordance with the typical guide for the care and use of laboratory animals (under the license of Institutional animal, medical ethics, bio-safety and bio-security committee (IAMEBBC) for experimentations on animal, human, microbes and living natural sources . No.33/320/IAMEBBC/IBSC).

2.2 Preparation of *N. naja* Venom

Cobra venom was collected from the Snake Rescue and conservation centre (SRCC) at Darusha, Rajshahi. In protein science lab, collected venom was milked, lyophilized and stored in a refrigerator at 4°C. Prior to use the venom was dissolved in 0.9% NaCl solution. LD_{50} dose of crude venom was determined by a technique previously described by Meier et al. [13]. The LD_{50} of cobra venom from our current study was calculated was 0.25 mg/kg of mice body weight.

2.3 Experimental Design

The animals were randomly distributed into two groups including control and envenomated group.

2.3.1 Control group

A total of five mice were injected intraperitoneally with only 200 μL physiological saline (0.9% NaCl) solution and serve as control.

2.3.2 Envenomated group

Twenty mice of the envenomated group were intraperitoneally injected with crude cobra venom as a single dose equivalent to LD_{50} of *Naja naja* venom (0.25 mg / Kg).

2.4 Histopathological Examination

Histopathological analysis of three organ from both groups of animal were conducted based on the technique previously described by Carleton et al. [14]. The animals were anesthetized using chloroform and sacrificed by cervical dislocation. Kidney, lung and intestine were collected and washed with normal saline, then fixed with Bouin's fluid (fixative) for -16-18 hour and subsequent washing under running tap water for one hour until complete removal of all traces of

chemical. Followed by washing, dehydration of the tissues was conducted by immersing the tissue in a series of gradually increasing concentrations of alcohol (50%, 70%, 80%, 95% and absolute alcohol) and embedded into paraffin wax for making blocks. The block was to be trimmed by removing of wax from the surface of block to expose the tissue. Sectioning of the tissue was performed by using a microtome (The microtome machine was sold from Tokyo, Japan by the trade name of SHIBUYA produced by optical. Co LTD). The microtome was pre-set to cut the tissue as thicknesses around 6 μm. Blocks Small ribbons of tissue sections were placed on microscopic slide with the help of warm distil water containing few drops of Mayer's albumen and deparaffinized with xylene solution. Haematoxylene and alcoholic eosin solution was used to stain the tissue for preparing permanent slide. Histopathological changes were observed under 20 x and 40 x magnification of a light microscope (The microscope was purchased from Italy by the trade name optika) and snaps were taken.

3. RESULTS AND DISCUSSION

3.1 Histopathological Changes in Kidney

Renal tissue of control mice (Fig. 1A) showed normal organization of the renal corpuscles with intact Bowman's capsules each enclosing by a tuft of glomerular capillaries. The typical renal tubule consists of intact proximal and distal convoluted tubules, loop of Henle and collecting tubules. Injection of cobra venom induced a range of histopathological alterations in renal tissues after 6 hours of the envenomation. Histological investigation of renal tissue showed that *N. naja* venom caused an inflammatory cellular infiltration, vacuolation in renal tubules, shrinking of glomeruli, degeneration of the distal and proximal convoluted tubules and raising space between the walls of Bowman's capsule (Fig. 1B). Since the toxic substances are circulated throughout the body with blood flow and whole blood samples of higher organism are filtered in the kidney. So kidney injury is among the common and most serious symptoms of cobra envenoming. The result of the present observations (Fig. 1B) were in agreement with findings reported by Amany et al. [15] which indicated that inflammatory cellular infiltration, vacuolation in the tubule and shrinkage of glomeruli in most cases in renal structure of envenoming mice injected with 1/2 LD_{50} *N. haje* venom.

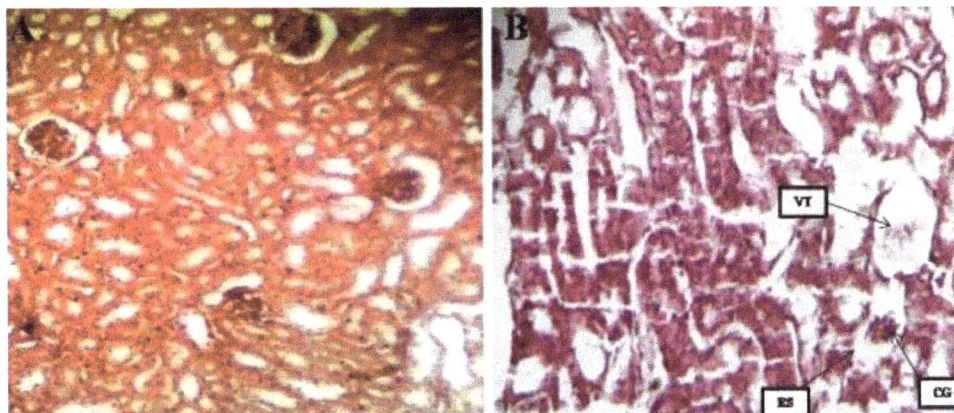

Fig. 1. Histological photograph of kidney tissue albino mice. (A) Control kidney showing compact glomerular and tubular structure. (B) Histological section from envenomated renal tissue showed vacuolation in the tubule (VT), shrinking glomeruli (CG) and raising space between the walls of Bowman's capsule (RS) .Sections were stained with hematoxylin and eosin (400X H and E)

According to Rahmy et al. [16] and Gunatilake et al. [17] a sub-lethal dose of the cobra venom was found to encourage injurious effects on the histological and histochemical patterns of the renal tissue of rabbit including complete necrosis in the glomeruli and proximal and distal convoluted tubular cell. Alterations in the glomerular architecture might be due to structural and molecular alterations in the glomerular basement membranes [18]. Renal tissue of envenomated animals showed tubular lesions with subsequent accumulation of inflammatory cells in the tubular tissues probably due to recovery of the injury by cobra itself. Cytoplasmic vacuolation appeared in renal tubules is a clinical sign of irregular lipid inclusions and fat metabolism happening under pathological cases [19]. Severe cellular swelling of the tubular epithelial lining cells resulting inflammation might be due to the action of venom phospholipase leading to cellular damage followed by scattering of the cellular contents in the tubular lumina [20]. The venom phospholipase (PLA_2) is probably the key factor responsible for tissue injury by disturbing cell membrane permeability through disorganizing of lipid bilayer on the plasma membrane resulting pore formation with subsequent influx of Na^+ and water [21]. Interaction between plasma membrane and phospholipase encourages the reduction of Na^+/K^+ ATPase activities with subsequent changes in the ionic gradients and followed by disordering the membrane lipid bilayer ultimately leading to cell death of envenomated person [22]. It is predicted that all the noticed renal injury

could be the reason of injecting higher Naja naja venom in the envenomed mice as formerly stated by Ismail et al. [23]. Renal injuries might be due to either direct or indirect effect of the venom composition [24]. The direct effects were mainly attributed to the exerting the venomous components which have an acute effects on the function as well as organization of the renal tissue, whereas the indirect action might be due to the deadly effect brought about by reactive metabolites or mediators produced in the kidney during envenoming [25].

3.2 Histopathological Changes in Lung Tissue

Pulmonary tissues of control mice (Fig. 2A) showed the normal and compact organization of bronchii, bronchioles and terminal bronchioles followed by specialized sac-like structures called alveoli consisting surface epithelium, blood vessels and supporting tissue surrounded by a double layer membrane structure called plura. Crude cobra venom persuaded some severe changes in their histological structure by showing significant inflammatory cellular infiltration and edema (Fig. 2B). The organism of envenomated group also showed alveolar haemorrhage and mionecrosis after 6 hours of envenomation with LD_{50} dose of cobra venom.

The essentiality to maintain the integrity of lung is inevitable for any higher animal due to its gaseous exchange activity and cleaning of blood cells with oxygen.

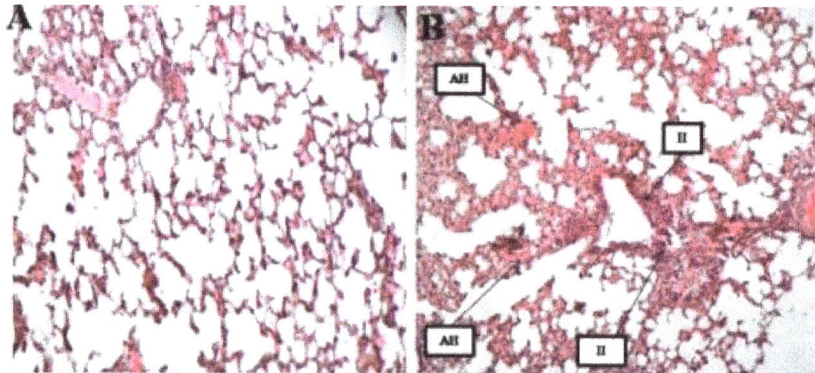

Fig. 2. Photomicrograph of histological section of lung tissue of albino mice. (A) The image from control pulmonary tissue showing compact configuration with intact alveoli and associated vessels and capillaries. (B) Lung tissue of envenomated group indicated extensive tissue damages and showing inflammatory cellular infiltration (II) and alveolar haemorrhage (AH). Sections were stained with hematoxylin and eosin (400 X H and E)

Individuals death occurred by cobra envenomation were mainly claimed due to the neurotoxic action of both presynaptic and postsynaptic neurotoxins followed by peripheral respiratory paralysis [26]. PLA_2 induces mionecrosis with subsequent infiltration of inflammatory cell to the affected tissue. The accumulation of polymorphonuclear leukocytes (mainly neutrophils) followed by macrophages indicated the establishment of neurotoxic effect of phospholipage [27]. In the current study, the rapid accumulation of polymorphonuclear cells suggesting severe attract of PLA_2 on the pulmonary tissue. Neutrophils have a significant contribution in the mediation of micro vascular damages leading to pulmonary tissue damage [28]. Leukocyte plays a key role in producing a range of inflammatory mediators contributing an important role in venom related respiratory failure. Inflammation may also be caused by reactive oxygen atom present in the venom and is also responsible for tissue damage (mionecrosis) occurring at the subsequent phases of inflammatory process. The lung integrity is maintained by the conservation of clear airspace and the regulation of fluid movement between the vascular, interstitial, and airway compartments. Edematous swelling of lung tissue with subsequent accumulation of pulmonary infiltrate blocks the alveolar airspaces disrupting gas exchange and lung mechanics, leading eventually to respiratory failure.

3.3 Histopathological Alteration in Intestinal Tissue

It was observed that the intestinal tissue from control mice was normally compact and regular structure of finger-like extensions villus, the mucosa and submucosa (Fig. 3A). But the tissue of envenomated grouped showed minor alteration in the structure (Fig. 4B). Histological evaluation of intestinal tissue demonstrated that *N. naja* envenomation caused no significant histopathological changes without cellular infiltration and mild hemorrhage (Fig. 3B). The degree of action of cobra venom on intestinal tissue was not sufficient to cause severe clinical sign compared with previously mentioned organs. Toxicological evaluation of *Naja naja* venom on intestinal tissue until precisely documented anywhere, though the toxin from other venomous animal like as scorpion have detrimental effect on intestinal tissues which showed slight bleeding (low dose), severe bleeding (high dose), mucosal necrosis and swelling and inflammation [29]. However, the current study demonstrated that mice were envenomated with cobra venom at LD_{50} producing a range of histopathological alterations such as inflammatory cellular infiltration in intestine probably due to recovery of the toxic effect by tissue itself and mild hemorrhage probably due to circulation of the venom components with blood flow in mucosa and sub mucosa (Fig. 3B).

The clinical injury, their sites and mode of action on tissue level along with the alteration in biochemical patterns must be taken in consideration in curing the victims bitten with *Naja naja* by producing the sustainable antivenom drugs for neutralizing the toxin reaction produced by venomous snake like as cobra.

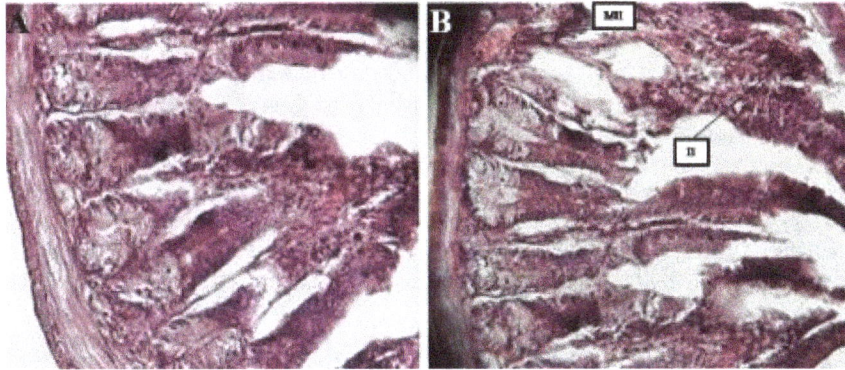

Fig. 3. Photomicrograph of histological section of intestinal tissue of albino mice (A) Intestinal tissue of control mice showed regular and compact configuration with villus, mucosa and sub mucosa. (B) Whereas, envenomated intestinal tissue appeared with inflammatory cellular infiltration (II) and mild hemorrhage (MH). Sections were stained with hematoxylin and eosin (400X H and E)

4. CONCLUSION

It can be concluded that the exposure of *Naja naja* venom encourages serious histopathological alterations on renal and pulmonary tissue and moderately on intestinal tissue of envenomed mice. Therefore, further studies need to be carried out for the isolation and purification of cobra venom and applying this valuable natural raw material in the discovery of anti-venom related drug along with other pharmaceutical valuable product.

COMPETING INTERESTS

Authors have declared that no competing interests exist.

REFERENCES

1. Swaroop S, Grab B. Snake bite mortality in the world. Bull World Health Organ. 1954;10:35–76.
2. Fry BG. From genome to venom, molecular origin and evolution of the snake venom proteome inferred from phylogenetic analysis of toxin sequences and related body proteins. Genome Res. 2005;15:403–420.
3. Rahmy TR, Hemmaid KZ. Histological and histochemical alterations in the liver following intramuscular injection with a sublethal dose of the Egyptian cobra venom. J Nat Toxins. 2000;9:21–32.
4. Al-Sadoon MK, Abdel Moneim AE, Diab MM, Bauomy AA. Hepatic and renal tissue damages induced by *Cerastes cerastes* gasperetti crude venom. Life Sci J. 2013;10:191–197.
5. Cherifi Fand Laraba-Djebari F. Isolated biomolecules of pharmacological interest in hemostasis from *Cerastes cerastes* venom. J Venom Anim Toxins Incl Trop Dis. 2013;19:11.
6. Tohamy AA, Mohamed AF, Abdel Moneim AE, Diab MSM. Biological effects of Najahaje crude venom on the hepatic and renal tissues of mice. J King Saud Univ Sci. 2014;26:205–212.
7. Yamazaki Y, Morita T. Snake venom components affecting blood coagulation and the vascular system: Structural similarities and marked diversity. Curr Pharm Des. 2007;13:2872–2886.
8. Westhoff G, Tzchatzsh K and Bleckmann H. The spitting behavior of two species of spitting cobras. J Comp Physio. 2005;191:873-881.
9. Cher CDN, Armugam A, Zhu YZ and Jeyaseelan K. Molecular basis of cardiotoxicity upon cobra envenomation. Cell Mol Life Sci. 2005;62:105–118.
10. El-Fiky MA. Hyperglycemic effect of a neurotoxic fraction (F3) from *Naja haje* venom: Role of hypothalamo-pituitary adrenal axis (HPA). J Nat Toxins. 1999;8:203–212.
11. Imam AH, Rahmy TR. Reactive astrocytic response and increased proliferatic cell nuclear antigen expression in cerebral cortex of envenomated rats. J Toxicol Toxin Rev. 2001;20:245–259.
12. Lougin M. Abdel Ghani, Mohamed F. El-Asmer, Osama A. Abbas, Tarek R. Rahmy.

Histological and immunohistochemical studies on the nephrotoxic effects of *Naja nigricollis* snake venom. Egyp J Nat Toxin. 2010;7:29-52.

13. Meier J, Theakston RD. Approximate LD_{50} determinations of snake venoms using eight to ten experimental animals. Toxicon. 1986;24:395–401.

14. Carleton HM, Drury RAB, Wallington EA. Carleton's histological technique. Ulster Med J. 1967;36(2):1.

15. Amany A. Tohamy, Aly F. Mohamed, Ahmed E. Abdel Moneim. Biological effects of *Naja haje* crude venom on the hepatic and renal tissues of mice. J Ki Saud Uni Sci. 2014;26:205–212.

16. Rahmy TR. Acute envenomation with cobra snake venom: Histological and histochemical studies on the kidney of rabbit. J Egypt Ger Soc Zool. 1999;28:79-104.

17. Gunatilake M, Jayakody RL, Angunawela P and de Tissera A. Direct nephrotoxic effects produced by venoms of Sri Lankan cobra, Russell's viper and hump nosed viper. Ceylon J Med Sci. 2003;46:61-66.

18. Miner JH, Sanes JR. Molecular and functional defects in kidneys of mice lacking collagen alpha 3 (IV)-implications for alport syndrome. J Cell Biol. 1996;135:1403-1413.

19. Ebaid H, Dkhil M, Danfour M, Tohamy A, Gabry M. Piroxicam induced hepatic and renal histopathological changes in mice. Lib J Med. 2007;2:82–89.

20. Rahmy TR. Histopathological, histochemical and biochemical studies on the effects of venoms of Sinai snakes on the animal body. Ph.D. Thesis, Zoology Department, Faculty of Science, Suez Canal University, Ismaila, Egypt; 1989.

21. Segelke BW, Nguyen D, Chee R, Xuong NH, Dennis EA. Structure of two novel crystal forms of *Naja naja* phospholipase A2 lacking C^{a2+} reveals trimeric packing. J Mol Biol. 1998;279:223–232.

22. Chethankumar M, Srinivas L. Gangliosides as potential inhibitors of *Naja naja* venom PLA2 (NV-PLA2) induced human erythrocyte membrane damage. Afr J Biochem Res. 2008;2:8–14.

23. Ismail M, Aly MHM, Abd-Elsalam MA, Morad AM. A three-compartment open pharmacokinetic model can explain variable toxicities of cobra venoms and their alpha toxins. Toxicon. 1996;34:1011-1026.

24. Martins AM, Monteiro HS, Junior EO, Mènezes DB and Fonteles MC. Effects of *Crotalus durissus cascavella* venom in the isolated rat kidney. Toxicon.1998;36:1441-1450.

25. Niesink RJM, de Vries J, Hollinger MA. Toxicology Principles and Applications. CRC Press, Inc. and Open University, Netherlands. 1996;453-725.

26. Nair X, Nettleton D, Clever D, Tramposch KM, Ghosh S, Franson RC. Swine as a model of skin inflammation. Phospholipase A2-induced inflammation. Inflammation. 1993;17:205–215.

27. Harris JB, Maltin CA. Myotoxic activity of the crude venoms and the principal neurotoxin, taipoxin, of Australian taipan, *Oxyuranus scutellatus*. Br J Pharmacol. 1982;76: 61–75.

28. Macnee W and Selby C. New perspectives on basic mechanisms in lung disease. 2. Neutrophil traffic in the lungs: Role of haemodynamics, cell adhesion, and deformability. Thorax. 1993;48:79–88.

29. Dehghani R, Khamehchian T, Vazirianzadeh B, Vatandoost H, Moravvej SA. Toxic effect of scorpion, *Hemiscorpius letpurs* venom on mice. The Journal of Animal & Plant Sciences. 2012;22:593-596.

15

Xylanases–from Microbial Origin to Industrial Application

Bushra Kalim[1], Nils Böhringer[2], Nazish Ali[1] and Till F. Schäberle[2*]

[1]Department of Zoology, Government College University, Lahore, Pakistan.
[2]Institute for Pharmaceutical Biology, University of Bonn, Bonn, Germany.

Authors' contributions

All authors wrote this review collectively and approved the final manuscript.

Editor(s):
(1) Marc Beauregard, Universite du Quebec a Trois-Rivieres, Canada.
(2) Giovanni DalCorso, Department of Biotechnology, University of Verona, Italy.
(3) Kuo-Kau Lee, Department of Aquaculture, National Taiwan Ocean University, Taiwan.
(4) Chung-Jen Chiang, Department of medical laboratory Science and Biotechnology, China Medical University, Taiwan.
Reviewers:
(1) Refaz Ahmad Dar, Biotechnology Division, IIIM ,Sanatnagar ,Srinagar, India.
(2) Gyula Oros, PPI HAS, Budapest, Hungary.

ABSTRACT

Xylanases are in the focus of research due to their potential to replace many current polluting chemical technologies by biochemical conversion. The field of application for xylanases is vast; it comprises industrial applications like wood pulp bio-bleaching, papermaking and bioethanol production. In addition, these enzymes can be applied as additives in food and beverage industry, and animal nutrition. However, considering the potential applications for these enzymes, the market share of xylanases is still low. Thus, the search for promising xylanases which tolerate relevant processing conditions and therefore can be used in industrial settings is an ongoing task. This review provides an overview of the enzymes reported from 2012 to mid 2014. Further, legal restrictions for the use of (genetically modified) organisms and enzymes are considered. This review provides an integrated perspective on the potential of specific xylanases for industrial applications.

Keywords: Enzymatic hydrolysis; GMO products; hemicelluloses; xylanase; xylo-oligosaccharides.

*Corresponding author: Email: till.schaeberle@uni-bonn.de

ABBREVIATIONS

AIA- advance informed agreement procedure; CBM- carbohydrate-binding module; CIBio- internal biosafety commission; CNBS- national biosecurity council; CTNBio- national biosecurity technical commission; ECF- elemental chlorine free; ECHA- european chemicals agency; EFSA- european food safety authority; EPA- US environmental protection agency; FDA- food and drug administration; FFDCA- federal food, drug and cosmetic act; GMO- genetic modified organism; GRAS- generally recognised as safe; MoA- ministry of agriculture; TCF- totally chlorine free; USDA- US department of agriculture; TSCA- toxic substances control act; XOS- xylo-oligosaccharides.

1. INTRODUCTION

Organisms use different strategies to obtain energy and basic building blocks for metabolism. Primary producers like plants utilise the energy from sunlight or inorganic substrates to produce organic molecules. Consumers instead feed on primary producers and other consumers. Finally, decomposers use remains of organisms and break it down to small organic molecules, *e.g.* sugars, amino acids and fatty acids. In contrast to dead animals the remains of plants are only poorly accessible due to their lignocellulosic cell wall. But, such plant material represents an enormous resource of organic matter and every year a new load is added to the environment from naturally and farmed sources. It comprises of cellulose, lignin and hemicelluloses, whereby from the latter the xylans are the most common ones. However, only a limited number of fungi and bacteria are able to degrade the plant cell wall into its monomers. These microorganisms are equipped with a set of hydrolytic enzymes to cleave and ultimately utilise the nutrients captured within these macromolecules. In contrast to the microorganisms, human used for a long time another way of exerting the energy bound in biomass, *i.e.*, direct burning. In many developing and underdeveloped countries burning wood still represents the most common energy source. However, at present times the usage of basic building blocks like sugars as basic chemicals for industry becomes more and more important. They have been used as starting material for the chemical and recently also biotechnological synthesis of various products, *e.g.* biofuel, bioplastics, and of high value fine chemicals. With evolving technology and commercial scale manufacturing processes, living standards have been increased over the last decades. However, many industrial production processes have a negative impact on the environment, due to the release of harmful and hazardous chemicals. Thus, to prevent persistent damages to nature, such as depletion of natural resources and environmental pollution, technologies have to become environment friendly.

A key to further development for the current polluting chemical technologies by biochemical conversion are enzymes, which are natural biocatalysts that perform under less extreme conditions than most chemical processes. Enzymes can play a vital role in chemical transformations since they offer advantages as compared to chemical (total) synthesis and degradation. They are very efficient, stereoselective, and mostly highly specific. Concerning plant biomass degradation xylanases (glycosidases which hydrolyse xylan) are of special interest, and these enzymes are largely used in various processes, *e.g.* bakery, distillery, diet of livestock, plant product processing, seed germination, and paper and pulp industry [1,2]. The molecular structure of xylanases and their mechanism of action are known [3]. Research on these enzymes had already been started in the early 50s as Sorensen [4] recorded the specificity and the products of xylanases activity from the fungus *Chaetomium globosum*. In the 80s it was claimed that the use of endoxylanases decreased the chemicals needed for bleaching Kraft pulp. Since then the production of industrial grade xylanases which are suitable for pulp industry – wherein they are used for pre-bleaching and deinking – represents the most important application for these enzymes. The global market of industrial enzymes in general is rapidly increasing. From one million dollars in 1990 it rose to 4.5 billion US$ in 2012. Current forecasts anticipate the global market to reach 7.1 billion by 2018. Because of their recognized potential as industrial biocatalysts, new xylanases are described in literature on a nearly weekly basis. This review provides an overview of the recently described enzymes by summarizing and comparing the reports of the last years (2012 – mid 2014).

2. XYLAN: AN OVERVIEW

The plant cell wall is made of three major polymeric components: (i) cellulose (polysaccharide fibres of 1,4-β-linked glucose units), (ii) lignin (complex polymeric structure of phenols) and (iii) hemicelluloses (polysaccharide fibres composed of glucose, mannans and xylan units).

After cellulose, xylan is the second most abundant polysaccharide in nature and accounts for one third of total renewable organic carbon on earth [5]. More than 10000 xylose units are polymerised by 1,4-β-linkages to form xylan strands (Fig. 1). Depending on the plant source, the xylans show structural variations. Different side chains can be attached to the linear backbone and based on their composition, xylans are grouped into four major families: (i) Arabinoxylans (AX) (side chains of single terminal units of α-L-arabinofuranosyl subunits) (ii) Glucoronoxylans (GX) (α-D-glucuronic acid or its 4-O-methyl ether derivative), (iii) Glucuroarabinoxylans (GAX) (α-D-glucuronic acid and α-L-arabinose) and (iv) galactoglucuronoarabinoxylans (GGAX) (characterised by the presence of β-D-galactopyranosyl residues on complex oligosaccharide side chains).

3. ENZYMATIC HYDROLYSIS OF XYLAN

Due to abovementioned structural and chemical variability in xylans, an array of specific hydrolytic enzymes is needed to break down these polymers to monomers, *i.e.* xylose residues. Complete xylan degradation is performed by a system of xylanolytic enzymes which catalyse several reactions: (i) Glycoside hydrolases, *i.e.* endo-1,4-β-xylanases and 1,4-β-xylosidase, which hydrolyse glycosidic bonds, (ii) Carbohydrate esterases, which hydrolyse ester bonds, and glucuronoyl esterases, and (iii) Polysaccharide hydrolases, *i.e.* α-L-arabinofuranosidase, α-galactosidase and α-4-O-methyl-D-glucuronidase, which are responsible to cleave off glycosidic bonds mainly in the chain substituent or side chains (Fig. 1). The cleavage of 1,4-β-glycosidic bonds in the xylan backbone by endoxylanases results in xylo-oligosaccharides (XOS) ranging from 2-10 monomers. Xylosidases cleave xylose residues from non-reducing termini of these short chain XOS. The other enzymes prepare the xylans for an effective hydrolysis by endoxylanases. Thus, carbohydrate esterases remove acetyl, feruloyl, and glucuronoyl groups. Acetylxylan esterases remove acetyl esters from O-2 or O-3 sites of backbone glycosyle units, whereas ferulic acid esterases deacetylate feruloyl groups from α-L-arabinoxylans. *p*-coumaric acid esterases hydrolyse ester linkages between arabinose side chain residues and *p*-coumaric acid. Further, glucuronoyl esterases demethylate O-6-methyl glucuronoyl in glucuronoarabinoxylans, and α-L-arabinofuranosidases remove α-L-arabinofuranosyl residues from non-reducing ends. α-galactosidases remove glycosidic bonds between galactose substituents and galactomannan. In hardwood xylan α-4-O-methyl-D-glucuronidases hydrolyse 1,2-α glycosidic bonds between D-glucuronic acids (ether 4-O-methyl-D-glucuronic acid) and xylose residues of XOS.

4. INDUSTRIAL APPLICATIONS OF XYLANASES

In 1819, Henry Braconnot had already reported the hydrolysis of cellulose into simple sugars by sulphuric acid [6]. Since then this procedure has been used for commercial scale hydrolysis of lignocellulosic biomass. However, acidic hydrolysis needs very high energy input as the reaction takes place at high temperature, and due to the low pH at which the hydrolysis takes place, the complete equipment needs to be acid resistant. These energy and cost consuming requirements reduce the cost effectiveness of the process. An alternative approach for the hydrolysis of cellulose and hemicelluloses is the use of suitable enzymes like xylanases. Dependent on the downstream following application xylanases are used to degrade the carbon-rich macromolecular fibres to different degrees. In biofuel production the xylan strands are degraded completely into fermentable sugars which subsequently can be used for fermentation and distillation of ethanol. For other applications longer xylan units (XOS) are the preferred compounds. A further field of application is the supplementation of products with xylanases to support downstream following biochemical reactions.

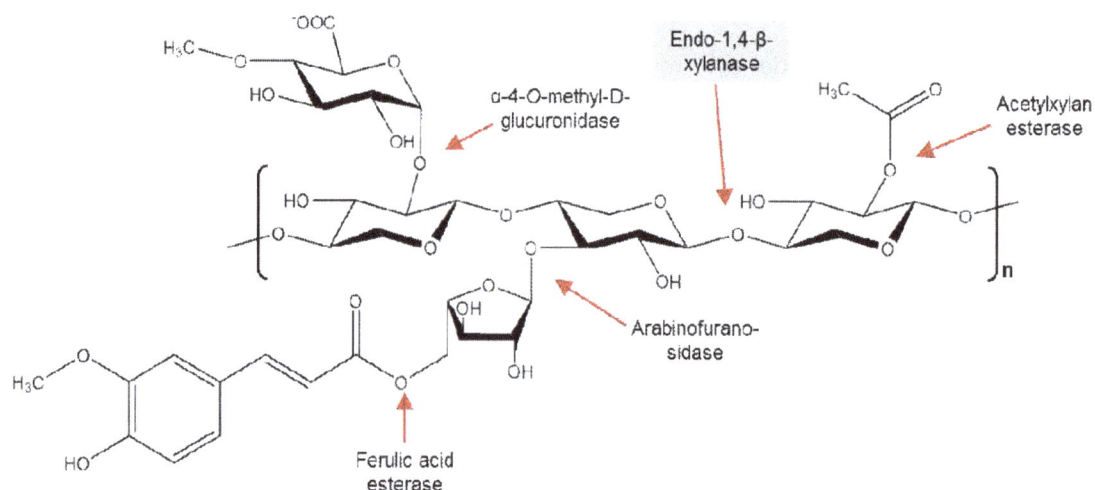

Fig. 1. Section of a xylan structure and the sites of its attack by xylanolytic enzymes
The backbone (in brackets) of the substrate is composed of 1,4-β-linked xylose residues. Theendo-1,4-β-xylanase, cleaving within the backbone, is highlighted in grey.

In addition, to produce xylanases for industrial applications, the easy access to the basic materials needed for their production is a clear advantage. For instance agricultural "waste", including crop harvest residue, is rich in lignocellulosic plant matter and can serve as raw material for xylanase production.

4.1 Bioethanol Production

Using plant biomass as starting material for the production of ethanol is a complex process which involves several conversions. First, the lignocellulosic biomass has to be delignified; yielding xylan and cellulose, which then can be hydrolysed to XOS and cellobiose, respectively. These oligosaccharides can subsequently be converted into simple sugars, *i.e.* xylose and glucose. The latter represent the compounds of interest for bioethanol production. Thus, here bifunctional xylanases, *i.e.* endo-1,4-β-xylanase with additional cellulase activity, can reduce the need for various enzymes [7]. This may help to make the saccharification process of lignocellulosic biomass more efficient and cheaper. A pretreatment of straw, using xylanases and cellulases can effectively improve downstream saccharification and fermentation, important for bioethanol production. Even though the biofuel technology continues to advance, the demand is stalling since the yields have to be increased and the costs decreased to compete with conventional fuels.

4.2 Animal Feedstock

Animal feed, including cereals like wheat, triticale and soy-based diet, are rich in lignocelluosic biomass. However, the viscous properties of lignocelluloses hinder their digestion. By enzyme supplementation to feedstock, nutrients entrapped in the macromolecules are released and thereby the digestion enzymeshave better access to their substrate. In that way the nutritional value of the diet increases. It was shown that enzymatic treatment of forages can improve the quality of silage, since it results in an improved rate of plant cell wall digestion in ruminants [8]. Addition of exogenous fibrolytic enzymes to the feedstock showed positive effects on animal growth by increasing the amount of fibres digested [9]. The *in trans* supplemented enzymes support the pre-digestion by breaking up the lignocellulosic biomass. This enables a more efficient use of the energy conserved in the sugar molecules. An additional effect of this pre-digestion is that the resulting sugars represent also nutritionally suitable compounds for the desired ruminal microflora, which in turn further assists digestion [10].

For more than fifty years the addition of enzymes has been known to improve animal performance, but their use as additives in animal feed was limited because it was assumed that enzymes will not survive proteolysis in the rumen. However, β-glucanases were introduced in poultry diet during the early 1980s in Scandinavia, and subsequently xylanases followed. Nowadays the 2014 world feed enzyme market is worth above 1 billion US$ and continues to increase by >5% annually. Positive effects of xylanases on animal growth were reported for fishery, piggery and poultry [11-13], as well as for cattle. Supplementation of fibrolytic enzymes like xylanases and cellulases significantly increased the average daily milk yield in Murrah buffaloes [8]. Similar results were obtained for goats [14]. The timing of the enzyme addition seems to be flexible, since an improved milk yield was also obtained by spraying forages with xylanases and cellulases from *Myceliopthora thermophilia* in appropriate doses just prior feeding [15].

However, it has to be noted that animal's response to enzyme supplementation may vary with its mode of implementation, *e.g.* whether the supplementation is applied to dry/fresh forage or silage or infused directly into the rumen, applied onto the complete diet or added to a component of the diet [16].

4.3 Production of Xylo-Oligosaccharides as Food Additive

Endoxylanase catalysed xylan hydrolysis results in XOS. The latter showed pre-biotic effects, and can be used in functional foods. They enhance the growth of potentially health promoting bifidobacteria in colon and thereby restrict the growth and proliferation of other harmful bacteria [17]. Therefore, XOS are used in beverages like soy milk, tea and coffee, dairy products, desserts like pastries, cakes, biscuits, puddings, jellies and jams, and special preparations for elderly people and children, or as active component of symbiotic preparations [17,18]. Further, XOS are reported to inhibit the growth of sarcoma-180 and other tumours, probably by indirectly stimulating the non-specific immunological host defence [19]. In addition to the immune-stimulating effects, anti-inflammatory activity was reported too [17]. These benefits make XOS interesting for the pharmaceutical industry.

4.4 Baking Industry

The use of enzyme preparations in bread making to produce fluffier bread has been reported long ago [20]. Xylans play an important role in bread making due to their water absorption capability and interaction with gluten. Utilization of xylanases improves bread qualities, *e.g.* good volume rise, smooth texture and appearance, and dry, balanced dough with prolonged shelf time. In 2012, xylanases, produced by a *Bacillus licheniformes* strain, were made commercially available by Novozyme as an additive for baking industry under the brand name "Panzea". Beside that one, different microbial sources for xylanases, which putatively can be applied in baking industry, were reported. The activity variation of these xylanases is given in the percentage of final loaf volume rise. Additional quantitative biochemical data are missing. Xylanase from *Aspergillus foetidus* was reported to increase the loaf volume by 53% [21]. Xylanases isolated from *Thermomyces maritime* and *Thermomyces lanuginosus* resulted in 60.3% rise [22]. An *Aspergillus oryzae* variant showed improved bread quality when compared to the xylanases produced by *Humicolain solens* and *Trichoderma reesei*.

4.5 Paper and Pulp Industry

Xylanases, as well as other enzymes, *e.g.* amylases, lipases and esterases, are applied in the paper and pulp industry. However, for their industrial application the xylanases has to be cellulose free, since the latter is the desired product of the paper industry. During the Kraft process, *i.e.*, the chemical conversion of wood into wood pulp (lignocellulosic fibrous material), the lignin-carbohydrate complex is hydrolysed. Thus, endo-1,4-β-xylanase can be used in this process to increase the extraction of lignin. In that way it becomes more accessible for bleaching. By this approach the consumption of chlorine chemicals for bleaching is reduced. In addition it was reported that, beside the positive effect on the environment, enzymatically treated pulp is brighter and has improved fibre quality. In 1979 it was observed that the incubation of unbleached kraft pulp with the fungus *Phanerochaete chrysosporium* resulted in delignification and in reduction of bleaching chemicals [23]. If this effect was due to xylanases alone or to the mix of enzymes produced by this strain was not further investigated. Later on, xylanase was isolated from the fungus *Schizophylum commune* by fractional precipitation and shown to have a significant hydrolytic activity against the hemicellulose content of the pulp [24]. Through the increasing portion of recycled paper used, xylanases gained further importance. This is due to the fact that such secondary (recycled) fibres require deinking to ensure a quality that could compete with that of virgin fibres.

4.6 Xylitol Production

Xylitol is a five carbon polyol with sweetness comparable to that of sucrose and is used in food and pharmaceutical industries. Several features foster its application in food industry, *e.g.* it is tolerable for diabetics, does not cause acid formation, has low viscosity, and has a cooling effect when dissolved in a solution [25]. Xylose can be reduced to xylitol in the presence of NADPH-dependant xylose reductase and thereby represent the substrate for xylitol production [26].

4.7 Fruit Juice and Beer Finishing

Xylanases together with many other hydrolytic enzymes like pectinases, amylases, tannases and carboxymethylcellulases are used in the brewing industry to reduce the turbidity of liquids and to clarify juices. Endo-1,4-β-xylanases help in juice extraction and filtration from vegetables and fruits by reducing the viscosity of the juices [27]. Further, they improve the yield of fermentable sugar from barley, which is useful in beer industry. Additionally, brewing liquid filterability is improved by the reduction of its viscosity [10]. Production of coffee, plant oils and starch represent further examples of application, since their extraction is facilitated by xylanases, which cleave the compact xylan layer of the seeds [28].

4.8 Degumming

In combination with a mild chemical pre-treatment, pectinases and xylanases play a vital role in the degumming and better separation of bast fibres. Enzymatic hydrolysis of bamboo materials make them suitable for textile processing [29]. Xylanases from *Bacillus subtilis* B10 were successfully used for degumming of ramie bast fibres [30]. Further, xylanases and pectinases are also used in the debarking of wood [28].

4.9 Seed Germination and Fruit Ripening

Germinating seeds naturally produce xylanases late in the germination process, which might help to access required nutrients for the better growth of the new plant. Xylanases are also reported to take active part in softening of fruits *e.g.* papaya. As the fruit starts to soften during ripening, endoxylanases play an important role by modifying the polysaccharides in the cell wall matrix [31]. Thus, xylanases might in the future play a role in the commercial scale ripening process of fruits. Many (exotic) fruits are usually harvested and transported in an unripe state and have to finish the ripening process in the country of destination.

5. MICROBIAL SOURCES OF XYLANASES

Xylanases are present in various microorganisms, *e.g.* free-living soil bacteria and fungi, or endosymbionts of ruminants and termites. Initially the enzymes have been identified by screening isolated environmental samples, *e.g.* axenic cultures, or heterogenic solutions of microorganisms, phenotypically for xylanase degrading activity. By the increasing number of available sequence information this process was speed up in the last years. Databases could be screened *in silico* to identify putative xylanase-encoding genes. Using molecular biological methods, *i.e.* cloning, enabled the overexpression of the target gene. By heterologous expression approaches even metagenomic environmental DNA samples became available for activity-based screenings. In Table 1 the most promising xylanases – judged by the activity reported under the given conditions – from different origin are given, and the top ten are described in detail. (A more comprehensive list can be found in the annex: Supplementary Table S1 and S2.)

B1 The highest xylanase activity reported from a bacterial host was 36633 IU/mg xylanase, using birchwood and oat spelt xylan as substrate. This was achieved using a purified xylanase originating from *Bacillus subtilis* CXJZ [32]. The enzyme is a weakly acidic, cellulase free xylanase with a molecular weight of 23.3 kDa and pI 9.63. The strain was obtained by screening ramie stems collected from the fields and subsequently partially submerged in natural river water. A genomic library of *B. subtilis* CXJZ was constructed and the resulting *E. coli* clones were screened for xylanolytic activity. From positive clones xylanase genes were amplified using non-degenerated primers. The sequence is deposited in GenBank under the accession number EU233656. The activity optima determined were 60°C and pH 5.8. The enzyme was stable up to 70°C and was easy to purify by ultracentrifugation and subsequent gel filtration chromatography. Further, it is reported that the strain is employed in the degumming lines of ramie textile mills in Hunan, China [32].

B2 A xylanase gene (*xyn11A*) coding a glycosyl hydrolase, a linker region and a carbohydrate-binding module (CBM) was cloned from the genomic library of alkalophilic *Bacillus* sp. SN5 [33]. The intact Xyn11A (366 amino acids) and Xyn11A-LC, the latter represents a truncated version without the CBM-linker, were expressed in *E. coli* BL21. Both purified proteins showed highest activity at 55°C. Optimal pH determined for Xyn11A was 7.5 and Xyn11A-LC showed a broad pH profile of 5.5-8.5 with >80% activity. Both enzymes were found to be alkali-tolerant and retained over 80% residual activity for 1 h after pre-incubation at pH 8.5-11.0 at 37°C. Xyn11A-LC was found to be more stable with lower affinity and catalytic activity using insoluble xylan as substrate but with higher specific activity for soluble beechwood xylan than Xyn11A. The presence of a CBM fused to Xyn11A implies that it might play an important role in degradation of insoluble xylan. Being alkali tolerant and cellulase free, Xyn11A-LC shows potential for applications in paper industry.

B3 An endoxylanase from *Bacillus brevis* ATCC 8246 showed activity in neutral to alkaline pH and stability at temperature of >55°C [34]. Enzyme production was carried out in liquid state cultures of *B. brevis* with wheat straw as main source of carbon. Using a combination of ammonium sulphate precipitation, gel filtration and ion exchange chromatography, the enzyme was purified to a 2.4 fold activity increase. The enzyme was stable between 45°C-90°C, and showed optimal activity at 55°C and pH 7.0. At 85°C the enzyme retained 50% of its activity. At 70°C its half-life was 2 h compared to 3 h at optimal temperature. High temperature tolerance of this *B. brevis* xylanase makes it a suitable candidate for Kraft pulp bio-bleaching.

B4 The bacterial strain *Paenibacillus macerans* IIPSP3A, producing a highly thermostable, halotolerant xylanase, was isolated from the gut of wood feeding termites using culture enrichment medium with beechwood xylan [35]. The purification of a xylanase with a molecular mass of 205 kDa and a pI of 5.38 by strong cation exchanger and gel filtration chromatography was reported [35]. Optimum pH

and temperature of the purified enzyme were 4.5 and 50°C, respectively. At 60°C the half-life was 6 h, and at 90°C the half-life was reported to be 2 h. High specific activity and stability at high temperature and pH make this enzyme suitable in bioethanol production and in paper and pulp industry as well.

B5 A thermostable bacterial xylanase was identified from *Bacillus halodurans* C-125 [36]. Xylanase C-125 was successfully cloned and expressed in *Pichia pastoris* and showed thermophilic and alkalophilic properties with an optimal activity at 70°C and pH 9.0. The enzyme was found to be cellulase-free. These properties make it compatible for applications in pulp and paper industry. Further, the enzyme enhanced brightness while preserving paper strength properties such as tensile index, tearing index, bursting index and post-colour number. C-125 XynA pre-treated wheat straw pulp allowed for a 10% decrease in chlorine consumption, maintaining brightness and kappa number at the same level.

B6 In a screening for strains producing thermostable xylanases, *Paenibacillus* sp. NF1 was isolated. Growing as a submerged culture, this strain showed high levels of extracellular xylanase activity [37]. The observed xylanase activity was 211.79 IU/mg. This value could be increased to 3081.05 IU/mg using oat spelt xylan as substrate after a 14.55-fold purification. The 37 kDa XynNF enzyme showed maximal activity at 60 °C and pH 6.0, respectively. Its optimal pH range is acidic, ranging from pH 4 to 7.

B7 A new xylanase Xyn11E was isolated and characterized from *Paenibacillus barcinonensis* BP-23 [38]. The complete sequence of the xylanase-encoding gene was obtained by gene walking on amplified DNA fragments. The xylanase was found to be dependent on *P. barcinonensis* BP-23 lipoprotein (LppX) for its expression in an active form using *E. coli* DH5α as heterologous host. Protein purification was carried out by immobilized metal affinity chromatography using a HisTrap HP column. The 20.7 kDa protein has a pI of 8.7. Optimal assay conditions were reported to be 50°C and pH 6.5. Enzyme activity on bleached pulps from different origin was also investigated. Xyn11E released reducing sugars from Elemental Chlorine Free (ECF) and Totally Chlorine Free (TCF) pulps from eucalyptus, sisal and flax, proving it to be suitable for enzymatic-assisted production of high cellulose-content pulps from paper-grade pulps.

B8 The hyperthermophilic eubacterium *Thermotoga petrophila* RKU 1 was isolated from an oil reservoir in Japan and was found to produce an endo-1,4-β-xylanase [39]. The strain was cultivated anaerobically in MMI medium. The xylanase encoding gene was amplified from the genomic DNA and heterologously expressed in *E. coli* strain BL21 Codon plus. The extracellular enzyme with a molecular mass of 40 kDa was purified by heat treatment, followed by anion and cation exchange chromatography. The optimal temperature and pH for this enzyme are 95°C and 6.0, respectively. The recombinant enzyme exhibited high stability with a half-life of 54.5 min at 96°C. The extremely high thermostability makes it a suitable candidate for application in paper and pulp industry.

B9 Another thermostable xylanase XynG1-1 was identified from *Paenibacillus campinasensis* G1-1 [40]. The bacterial strain was isolated from cotton stalk stockpile samples and cultivated in complex medium for xylanase production. A xylanase gene was amplified from the genomic DNA, subsequently cloned and expressed in *E. coli* BL21 (DE3). The protein with a mass of 41.3 kDa was purified to homogeneity by hydrophobic interaction, gel filtration, and ion exchange chromatography. Optimal conditions for the cellulase-free XynG1-1 were determined to be 60°C and pH 7.5, respectively. Stability was preserved between 70°C-80°C and over a wide pH range of 5.0-9.0. The authors state that the enzyme should be greatly valuable in various industrial applications, especially for pulp bleaching pre-treatment.

B10 A putative endo-1,4-β-xylanase (XylB8) was identified by sequence analysis of cloned metagenomic DNA from termite gut [41]. An actinobacterial xylanase gene was sequenced, cloned and heterologously expressed in *E. coli*. The recombinant protein was purified by affinity chromatography and its specificity appeared to be limited to xylan. Maximal activity was determined at 55°C and pH 5.0. The enzyme was reported to have potential application in animal feed, textile and paper industries.

F1 A promising xylanase (XynR8) was cloned from un-purified rumen fungal cultures. Later on *Neocallimastix* sp. was suggested as the real origin [42]. Heterologous expression of XynR8 in

E. coli resulted in a highly active (66672 U/mg) and stable (pH 3-11) protein. The optimal temperature was 55°C. By inserting utations N41D and N58D, which according to the authors should stabilize two loops within the protein structure, the tolerance to pH 2 was increased by 24.6%. In addition, the activity was increased 1.4 fold for N58D (96689 U/mg) and 1.1 fold for N41D (79645 U/mg) compared to the wild type, putting these xylanases among the most active onestowards soluble oat spelt xylan. The activity of this enzyme at very low pH makes it an interesting candidate in biofuel production.

F2 The highest xylanase activity from crude extract was obtained by using a gene originating from *Aspergillus niger* IME-216, which was overexpressed in an industrial strain of *Saccharomyces cerevisiae* [43]. Using cDNA as template a fusion fragment consisting of phosphoglycerate kinase promoter, α-factor signal peptide, xylanase, and a terminator was constructed. This construct was integrated into the genome of *S. cerevisiae* YS2. The xylanase production of the transgenic yeast was 1.5 fold higher (74800 U/mL) than the value reached by the host strain under flask culture at 28°C for 72 h. The authors speculate about the use of recombinant *S. cerevisiae* YS2 strains in bioethanol production by simultaneous fermentation of glucose and xylose, even though the 1.5 fold increase in a transgenic host is only slightly above the wild type strain.

F3 *Paecilomyces thermophila* xylanase (XynA) was functionally expressed in *Pichia pastoris* GS115 [44]. Optimal activity of the recombinant XynA was observed at 75°C and neutral pH with 30 min stability recorded at 80°C. The high activity (52940 U/mg) and the robust expression in yeast make PtxynA worthy of further investigations.

F4 The putative xylanase gene *xyl10* from *Aspergillus niger* was codon-optimised for *P. pastoris* and synthesized [45]. The final gene product was inserted in *P. pastoris* X33 under the control of the glyceraldehyde-3-phosphate dehydrogenase (GAP) promoter. Purification of the recombinant protein was carried out by size exclusion chromatography. Recombinant Xyl10 showed optimal activity at 60°C and pH 5.0. The enzyme maintained 90% of its original activity at 60°C for 30 min at pH 5.0, and around 74% after incubation at pH 3.0-13.0 for 2 h at 25°C. Recombinant Xyl10 exhibits great potential in various industrial applications, due to its high activity (32000 U/mg).

F5 *Aspergillus usamii* E001 was used as donor of a xylanase gene encoding AuXyn11A [46]. The N-terminal amino acids were replaced by the corresponding ones from the hyper thermostable xylanase Syxyn11. The *auXyn11A* part was amplified from total RNA whereas the *syXyn11* part was synthesized. Both genes were transferred to *E.coli* JM109, serving as parent genes for the construction of a hybrid xylanase gene, *AExynM*. AuXyn11A and AEXynM were expressed in *P. pastoris* GS115. A combination of ammonium sulphate precipitation, ultrafiltration, and gel filtration was used for purification. Optimum temperature reached 75°C for the thermo-stabilised version, compared to 50°C for AuXyn11A. Further, three amino acids in the replaced N-terminus were tested and shown to contribute to high thermal stability.

F6 The fungus *Trichoderma reesei* is an established host for expression of homologous or heterologous genes. Xylanase II was homologously overexpressed in *T. reesei* QM9414 by putting the gene under the control of the promoters of *pdc* (encoding pyruvate decarboxylase) and *eno* (encoding enolase) [47]. By real-time reverse-transcription polymerase chain reaction these promoters were shown before to be extremely active under high glucose conditions. The recombinant strains produced 9266 U/ml (*pdc* promoter) and 8866 U/ml (*eno* promoter) of xylanase activities in the culture supernatant. Xylanase II accounted for >80% of the total protein secreted by the recombinant strains, indicating the high potential for rational promoter screening for optimising homologous and heterologous xylanase production.

F7 A 26 kDa endoxylanase was obtained from the wood-rotting fungus *Schizophyllum commune* [48]. The corresponding gene was cloned and expressed with a C-terminal His$_6$-tag in *P. pastoris* GS115. Following Ni-NTA affinity purification, optimal conditions for activity were determined to be 50°C and pH 5. Enzyme activity improved in the presence of cations like potassium, sodium, lithium, cadmium and cobalt whereas EDTA, mercury and ferric ions inhibited its activity. It was suggested that this enzyme might serve as a major component of a lignocellulosic degrading enzyme cocktail.

F8 The activity of an extracellular endoxylanase PoXyn2 from the filamentous fungus *Penicillium occitanis* Po16 was reported to be 2368 U/mg in an assay using oat spelt xylan as substrate. The cDNA was cloned in an expression vector that allowed to fuse a N-terminal His$_6$-tag to the enzyme. The construct was integrated in *P. pastoris* X-33 under the control of the constitutive GAP promoter. Recombinant protein was expressed extracellular and purified to its homogeneity by using immobilized metal affinity chromatography (Ni-NTA resin). By using *P. pastoris* X33 as heterologous host the activity towards oat spelt xylan was increased to 8550 U/mg [49]. The 30 kDa enzyme showed optimal activity at 50°C and pH 3. Ability to work under acidic pH optimum and wide pH stability allow to predict potential applications of PoXyn2 in feed and food industries.

F9 Two extracellular xylanases, designated Xyn-1 and Xyn-2, were isolated from *Rhizopus oryzae* NRRL 29086, a strain previously isolated from dew-retted flax [50]. The two proteins Xyn-1 (32 kDa) and Xyn-2 (22 kDa) were expressed in *E.coli* and characterised after purification to electrophoretic homogeneity. The activities determined were 605 U/mg and 7710 U/mg, respectively. Melting temperature of these enzymes was estimated to fall in the range of 49.5-53.7°C. Xyn-1 and Xyn-2 demonstrated release of xylose from flax shives-derived hemicelluloses as model feed-stock.

F10 A 20 kDa heat resistant xylanase was isolated from *Aspergillus niger* SCTCC 400264 (Li 2014) [51]. The corresponding *xynB* gene was transferred and expressed as His-tagged protein in *E. coli* as well as in *P. pastoris*, whereby enzyme production was four times more important in *P. pastoris*. Maximal enzyme activity was recorded to be 4757 U/mg under optimal conditions, *i.e.* 55°C and pH 5.0. The enzyme showed 86% residual activity after 10 min heat treatment at 80°C (74% after 30 min). The gene integration was stable, and after 50 generations the cells were tested positive for the *xynB* gene. This xylanase could be used in animal feed applications, since it has 82% residual activity at body temperature, and can preserve >50% relative activity under acidic pH. Further, the enzyme is thermostable, a feature necessary for high-temperature requirements of typical feed pelleting process.

Table 1. Xylanases of bacterial and fungal origin.[a]

Strain number[b]	Natural producer	Heterologous host	Enzyme purification	Xylanase activity [IU/mg]	Used substrate	Optimal conditions	Ref.
B1	*Bacillus subtilis*	*E.coli* DH5α	Purified	36633.4	Birchwood and oat spelt xylan	60°C pH 5.8	[32]
B2	*Bacillus* sp. SN5	*E. coli* BL21 (DE3)	Purified	4511.9	Beechwood xylan	55°C pH 7.5	[33]
B3	*Bacillus brevis* ATCC 8246	—	Purified	4380	Agro-waste *e.g.* Wheat straw	55°C pH 7.0	[34]
B4	*Paenibacillus macerans* IIPSP3	—	Purified	4170±23.5	Beechwood xylan	60°C pH 4.5	[35]
B5	*Bacillus halodurans* C-125	*Pichia pastoris*	Crude	3361 U/ml	Wheat Straw Pulp	70°C pH 9.0	[36]
B6	*Paenibacillus* sp. NF1	—	Purified	3081.05±4.12 3045.93±4.34 2533±5.66 2460±6.74	Oatspelt Xylan Birchwood Xylan Rice hull Xylan Bran Xylan	60°C pH 6.0	[37]
B7	*Paenibacillus barcinonensis* BP-23	*E. coli*	Purified	3,023	Beechwood xylan	50°C pH 6.5	[38]
B8	*Thermotoga petrophila* RKU 1	*E.coli* BL21	Purified	2600 1655	Birchwood xylan Beechwood xylan	96°C pH 6.0	[39]
B9	*Paenibacillus campinensis* G1-1	*E.coli* BL21 (DE3)	Purified	1865.5	Birchwood xylan	60°C pH 7.5	[40]
B10	*Actinobacterium*	*E.coli*	Purified	1837	Xylan	55°C pH 5.0	[41]
B11	Microbial consortium	*E.coli* DE3	Purified	1794.4±136.6 1029.5±55.2	Birchwood xylan Beechwood xylan	65°C-70°C pH 9.0-10	[52]
B12	*Bacillus pumilus* SSP-34	—	Purified	1723	Oat spelts xylan	50°C pH 6.0	[53]
B13	*Gracilibacillus* sp. TSCPVG	—	Purified	1667	Birchwood xylan	55°C pH 7.5	[54]
B14	*Thermobacillus*	*E.coli* JM109 (DE3)	Purified	1170	Birchwood xylan	60°C	[55]

Strain number[b]	Natural producer	Heterologous host	Enzyme purification	Xylanase activity [IU/mg]	Used substrate	Optimal conditions	Ref.
B15	xylanilyticus Rumen microbiota of Hu sheep	E.coli BL21 (DE3)	Purified	1150	Birchwood xylan	pH 6.2 50°C	[56]
B16	Bacillus sp. QH14	E.coli BL21 (DE3)	Purified	Wild type: 1148.56 E. coli: 700.47	n.g	pH 5.0 60°C pH 9.2	[57]
B17	Cellulosimicrobium sp. MTCC 10645	—	Purified	960	Birchwood xylan	50°C pH 7.0	[58]
B18	Streptomyces sp. Strain S9	Pichia pastoris	Purified	950.8+23.12	Birchwood xylan	70°C pH6.5	[59]
B19	Cellulosimicrobium sp. Strain HY-13	—	Crude	788.3	Birchwood xylan	45°C pH 6.0	[60]
B20	Bacillus amyloliquefaciens CH51	E.coli BL21	Purified	701.1	Birchwood xylan	25°C pH 4.0	[61]
F1	Neocallimastrix sp.	E.coli	Purified	N58D: 96689 N41D: 79645 Wild type: 66672	soluble oat spelt xylan	55°C pH 2.0	[42]
F2	Aspergillus niger IME-216	Saccharomyces cerevisiae YS2	Crude	74800 U/ml	Xylan	28°C	[43]
F3	Paecilomyces thermophile	Pichia pastoris GS115	Purified	52940	Birchwood xylan	n.g. 75°C pH 7.0	[44]
F4	Aspergillus niger	Pichia pastoris X33	Purified	32000	Beechwood xylan	60°C pH 5.0	[45]
F5	Aspergillus usamii E001	Pichia pastoris GS115	Purified	AuXyn11A: 9266 AeXynM: 8866	Corncob xylan	50 & 70C pH 4.6	[46]
F6	Trichoderma reesei QM9414	E. coli Top10F'	Crude	P_{pdc}: 9266 P_{eno}: 8866	Birchwood xylan	50°C pH 5.3	[47]
F7	Schizophyllum commune	Pichia pastoris GS115	Purified	9000	Beechwood xylan	50°C pH 5.0	[48]
F8	Penicillium occitanis Pol6	Pichia pastoris X-33	Purified	8549.85	Oat spelt xylan	50°C pH 3.0	[49]
F9	Rizopus oryzae strainNRRL 29086	E. coli	Purified	Xyn2: 7710 Xyn1: 605	Birchwood xylan	35°C pH 5.0	[50]
F10	Aspergillus niger SCTCC	Pichia pastoris	Crude	4757	Oat spelt xylan	80°C	[51]

Strain number[b]	Natural producer	Heterologous host	Enzyme purification	Xylanase activity [IU/mg]	Used substrate	Optimal conditions	Ref.
	400264						
F11	Chaetomium globosum	E.coli	Purified	4530+230	Rice straw	pH 4.0-6.0 40°C	[62]
F12	Armillaria gemina SKU2114	—	Crude	1270 U/ml	Rice straw	pH 5.5 n.g.	[63]
F13	Armillaria gemina	Trichoderma reesei Qm9414	Purified	1270 U/ml	Birchwood xylan	n.g.	[64]
F14	Penicillium oxalicum B3-11		Purified	1856+53.5	Beechwood xylan	50°C pH 5.0	[65]
F15	Aspergillus usamii E001	Pichia pastoris GS115	Purified	912.6 U/ml	Insoluble Corncob xylan	50°C pH 5.0	[66]
F16	Aspergillus niger DSM 1957	Pichia pastoris GS115	Purified	808.5	Birchwood xylan	50°C pH 7.0	[67]
F17	Gloephyllum trabeum	Pichia pastoris GS115	Purified	727.2+15.75	Beechwood Xylan	70°C pH 7.0	[68]

[a]The best xylanases (activity >700 U/mg) reported in recent years are given; [b]bacterial hosts are numbered with B and fungal hosts with F; n.g.: not given; Ref.: reference

6. LEGAL RESTRICTIONS FORTHE INDUSTRIAL USE OF XYLANASES

Additionally to scientific issues related to xylanase research, it has to be kept in mind that legal regulations influence research and application. While the application in industrial processes like pulp bleaching seems rather unproblematic, legal issues are obviously at stake for enzymes used in the food chain. For the application of xylanases in non-food related industrial processes, general product safety regulations and directives for genetic modified organisms (GMOs) safety which differ from country to country, have to be taken into account. When used in terms of food processing or feeding, additionally food safety regulations have to be considered.

Researchers and companies working with wild type strains have the advantage that for the final product as well as for the production process, no GMO regulations have to be taken into account. Independently from the fact that the whole crude extract or purified enzymes are used. However, wild type strains may have several disadvantages, e.g. slow growth rates, low yield, or special growth requirements, which hinder their application due to economic reasons. Since most of these wild type strains are fungi, special cultivation methods like solid state fermentation have to be used for large scale cultivation, an important disadvantage over classical liquid fermentation methods. Another disadvantage is that depending on the process, e.g. the bleaching of pulp, only xylanase activity is desired, while the wild type strains might also express other hydrolytic enzymes such as cellulases. Further, enzymatic degradation of lignocellulose demands the interplay between a whole set of different hydrolytic enzymes, that are usually not expressed in one organism providing all activities required. Therefore, heterologous expression of xylanase genes has advantages, e.g. vast yield enhancement by (heterologous) overexpression in host strains with optimised cultivation methods, and access to proteins encoded in the genome of so far unculturable microorganisms, solely detected in metagenomics approaches using environmental DNA samples.

The handling of GMOs and GMO products is essentially regulated by two international agreements as well as the respective national laws. In the following, a short overview of the international agreements and the statutory regulations in Brazil, China, EU and USA is given.

International Regulations The "Cartagena Protocol on Biosafety to the Convention on Biological Diversity" regulates the handling of living genetically modified organisms in terms of trade between states. This protocol is an addition to the "Convention on Biological Diversity" from 1993, it has been signed by 166 countries and became valid in 2003 [69,70]. The basic claim of the protocol is that the importing country's environment and the population's health have to be protected from the possible risks posed by the uncontrolled release of GMOs. Therefore, the importing countries have to get access to all scientific information about the GMO allowing to make an informed decision regarding whether an import is allowed or not. This principle is referred to as "Advance Informed Agreement Procedure" (AIA). Particularly excluded from the AIA are GMOs for pharmaceutical use and for food, feeding and respective processing purposes. Before using GMOs it shall be ensured that the organism underwent some monitoring processes that correlate with the respective generation time. In addition, prior to an intentional release of GMOs into the environment, risk assessments have to be performed by the particular party [71]. The "Nagoya – Kuala Lumpur Supplementary Protocol on Liability and Redress to the Cartagena Protocol on Biosafety" (Supplementary Protocol) is an addition to the original Cartagena Protocol. This protocol regulates the responsibilities in case of damages caused by GMOs. This includes the damage by authorised releases as well as unintentional or illegal trans-boundary movement [72].

Brazil is the second largest producer of genetically modified plants for livestock feeding, and one of the biggest biofuel producers worldwide. The Brazilian constitution defines the intact environment as public good, necessary for a healthy lifestyle and following the precautionary principle, the biosafety law (Law No. 11,105) [73] which sets safety regulations for the handling of GMOs and GMO products was enforced by the congress. It distinguishes between laboratory (development and testing) and commercial (production and processing) use of GMOs. According to this law, the "National Biosecurity Technical Commission" (CTNBio) has to set standards for the handling of GMOs, e.g. safety guidelines. This central authority reviews all biotechnological projects and decides whether to approve them or not case by case. The "National

Biosecurity Council" (CNBS) has to approve, and if necessary to create administrative regulations for, the introduction of GMOs and GMO products considering social, economic, and national interest. Additionally, each institution dealing with the creation, production or research of GMOs has to set up an "Internal Biosafety Commission" (CIBio). Thus, the release of GMOs has to be approved by the CTNBio, as well as the CNBS if the release is for commercial purposes. Food and feed containing GMOs or GMO products have to be labelled according to the ministerial act No. 2.658 if the GMO content is higher than 1% [74]. Recombinant proteins are considered GMO products. Therefore, research and production of xylanases, if obtained from transgenic strains, have to be approved by the CTNBio in each case and have further to be reviewed by the CNBS if a product, containing recombinant xylanases, enters the market.

China´s legislation on GMOs mainly focuses on genetically engineered organisms for agricultural use, including plants, animals or microorganisms. GMOs underlie the "Regulations on Administration of Agricultural Genetically Modified Organisms Safety" [75]. According to the regulations, research and testing of GMOs has to take place in facilities that have the standard for the biosafety level required by the organism investigated. If this involves organisms requiring biosafety level 3 or 4, the Ministry of Agriculture (MoA) has to be informed before the start of the project. New products derived from GMOs need a so called GMO Safety Certificate, issued by the MoA after three stages of product testing. GMOs and GMO products for food and feeding purposes can be approved by the Ministry of Health after risk assessment [76]. The regulations on GM food also include a labelling system and a tracing system, which forces producers of GM food to keep tracking files of their products at least for two years and to label products containing GMOs and their products accordingly. Since no further legislation concerning GMOs has been enforced yet, it is not possible to make specific statements about the situation with xylanases in China. While the food, drug and agricultural sectors are well regulated, it is however not clear if recombinant enzymes are considered as GMO products or as chemical substances. Additionally, it seems that there are no regulations for non-food, non-drug or non-agricultural biotechnology applications other than basic environment protection laws.

EU authorities have legislative competences for all 28 members, especially for healthcare, consumer and environment protection [77]. The EU directives represent a minimum standard that has to be fulfilled for products to be produced and marketed within the EU. While each of the 28 countries can deny products that fulfil the EU standards, it is not possible to loosen the national regulations below the EU standards. To protect the environment and the population's health, these EU standards follow the precautionary principle of the Cartagena Protocol. Essentially three directives regulate the handling of GMOs and GMO products within the EU. The directives as such have to be implemented into national laws by EU members, determining which authorities are in charge of control. Directive 1907/2006 determines that all substances imported into or manufactured in the EU in amount over one ton have to be registered by the "European Chemicals Agency" (ECHA) [78]. The registration requires, among others detailed information about the manufacturer, the identity of the substance and safety guidelines. Particularly excluded are substances for food and feeding purposes. Directive 2009/41/EG regulates the handling of GMOs within closed systems [79]. These regulations apply to every production process that includes transgenic microorganisms and states that all appropriate measures have to be taken to avoid damage to the environment and/or the public health. This includes measures to ensure that no living organisms escape the closed system. Directive 2001/18/EC regulates the intentional release of GMOs into the environment [80]. This also applies to production methods where the remaining of the GMO within a closed system cannot be ensured. Before the release of a GMO into the environment, a case by case environmental impact assessment has to be performed. This includes practical experiments to ensure the safety of the organism in the respective environment and requires detailed knowledge about the host organism, the implemented gene, the used vector, the application and processing and potential health and environmental threats. Toxic products and potential allergens (each new protein) are also recognised as potential health treats. The directive 1829/2003 regulates the utilisation of GMOs and GMO derived products in food and feeding [81]. A GMO or GMO product that is utilised in food and feeding has to be approved by the "European Food Safety Authority" (EFSA). Additionally, it has to be ensured that all ingredients can be traced back to the producer and that products containing GMOs or GMO products are labelled as such. Labelling is not necessary if the GMO content is unintended but inevitable due to the process conditions and lies below 0.9% of the product. This would apply to purified (whether GMO or non GMO derived) enzymes, which have to be approved by the EFSA as ingredients in food or feeding products. This is especially important, since natural producers, *e.g.* fungal strains, are known to produce toxins. The dilution or even the removal of toxins by technical processes is forbidden. To bypass the requirement to label the product as GMO product, it is also necessary to purify the xylanase at least below the previously mentioned threshold. This is of great interest, since the consumer acceptance of GMO products is very low in certain EU member states.

USA, as the largest producer of GMO products, has no particular federal laws regulating the use and release of genetically modified organisms or their products. Instead, the regulations relevant to the particular product are applied. Therefore, the products are categorised as agricultural, food and drug and as general chemical products. Depending on the product, it has to undergo the approval procedures of one or more of the institutions. These regulations focus on the safety of the end product rather than the production process. The authorities in charge are the US Department of Agriculture (USDA), the Food and Drug Administration (FDA) and the US Environmental Protection Agency (EPA) [82]. Within this scheme genetically modified microorganisms for non-food and non-healthcare application are generally considered as chemical substances by the Toxic Substances Control Act (TSCA) [83] and their safety has to be approved by the EPA. Products derived from GMOs for non-pharmaceutical, non-food, feeding or processing purposes also have to be approved by the EPA unless they are listed in the TSCA Chemical Substance Inventory [84]. Microorganisms and derived products for pharmaceutical use or use in food, feeding or processing, however, are under the responsibility of the Federal Food, Drug, and Cosmetic Act (FFDCA) or the Public Health Service Act (PHSA) and have to be approved by the FDA. Considering this, strains for recombinant production of xylanases at industrial scale within the USA have to be approved by the EPA while the production of xylanases using wild type strains does not need any approval. The product (endo-1,4-β-xylanase (CAS-Nr: 9025-57-4)) also does not have to be approved by the EPA, since it is already listed on the TSCA Chemical Substance Inventory. Xylanases for application in food, feeding and processing are under the responsibility of the FFDCA, regardless of the production process. The FDA however approves (endo-1,4-β-)xylanases as food and feeding additives as well as for food processing as "generally recognised as safe" (GRAS) [85]. Nevertheless, novel formulations and mixed enzyme formulations have to be approved by the FDA.

7. CONCLUSION

Although the use of wild type strains is not as strictly regulated as the use of GMOs and GMO products, it seems that only the optimisation of microbial strains renders them, or their products, promising for industrial applications. Enzymes identified by screening various habitats show different qualities. The exhaustive analysis of existing data will promote the rational design of new enzymes with the desired properties. Industrial producer strains will form the basis for heterologous hosts. Progress in biotechnology will unleash the potential of xylanases in the next years.

ACKNOWLEDGEMENTS

Funding by the Higher Education Commission of Pakistan to B.K. and N.A. is greatly acknowledged.

COMPETING INTERESTS

Authors have declared that no competing interests exist.

REFERENCES

1.   Harris AD, Ramalingam C. Xylanases and its application in food industry: A review. JES. 2010;1:1-11.

2.   Goswami GK, Pathak RR. Microbial xylanases and their biomedical applications: A review. Int J Basic Clin Pharmacol. 2013;2:237-46.

3.   Ratanakhanokchai K, Waeonukul R, Pason P, Tachaapaikoon C, Kyu KL, Sakka K, etc. *Paenibacillus curdlanolyticus* Strain B-6 multienzyme complex: A novel system for biomass utilization. In: "Biomass Now - Cultivation and Utilization", edited by Miodrag Darko Matovic; 2013. ISBN 978-953-51-1106-1.

4. Sorensen H. On the specificity and products of action of xylanases from *Chaetomium globosum Kunze*. Physiol Plant. 1952;5:183-97.

5. Collins T, Gerday C, Feller G. Xylanases, xylanase families and extremophilic xylanases. FEMS Microbiol Rev. 2005;29:3-23.

6. Braconnot H. On the conversion of the timber body gum, sugar and an acid of a particular nature by means of sulphuric acid; conversion of the woody substance by potash. Ann Chim et Phys. 1819;12:172-95.

7. Sharma M, Kumar A. Xylanases: An overview. BBJ. 2013;3:1-28.

8. Shekhar C, Thakur SS, Shelke SK. Effect of exogenous fibrolytic enzymes supplementation on milk production and nutrient utilization in Murrah buffaloes. Trop Anim Health Prod. 2010;42:1465-70.

9. Beauchemin KA, Colombatto D, Morgavi DP, Yang WZ. Use of exogenous fibrolytic enzymes to improve feed utilization by ruminants. J Anim Sci. 2003;81(E. Suppl. 2):E37-E47.

10. Garg N, Mahatman KK, Kumar A. Xylanase: Applications and Biotechnological Aspects. Lambert Academic Publishing AG & Co. KG, Germany; 2010.

11. Huichang C. Application of Xylanase in Fish Feed. J Anhui Agricult Sci. 2006;34:3077.

12. Malagutti L, Colombini S, Rapetti L, Pirondini M, Magistrelli D, Galassi G. Xylanase and benzoic acid in the fattening heavy pig: effects on growth performance and on nitrogen and energy balance. Energy and Protein Metabolism and Nutrition. 2010;403-4. ISBN: 978-90-8686-153-8.

13. O'Neill HV, Mathis G, Lumpkins BS, Bedford MR.The effect of reduced calorie diets, with and without fat, and the use of xylanase on performance characteristics of broilers between 0 and 42 days. Poult Sci. 2012;91:1356-60.

14. Bala P, Malik R, Srinivas B. Effect of fortifying concentrate supplement with fibrolytic enzymes on nutrient utilization, milk yield and composition in lactating goats. Anim Sci J. 2009;80:265-72.

15. Beg QK, Kapoor M, Mahajan L, Hoondal GS. Microbial xylanases and their industrial applications: A review. Appl Microbiol Biotechnol. 2001;56:326-38.

16. Yang WZ, Beauchemin KA, Rode LM. A comparison of methods of adding fibrolytic enzymes to lactating cow diets. J Dairy Sci. 2000;83:2512-20.

17. Aachary AA, Prapulla SG. Xylooligosaccharides (XOS) as an emerging prebiotic: Microbial synthesis, utilization, structural characterization, bioactive properties and applications. Compr Rev Food Sci F. 2011;10:1-15.

18. Kumar GP, Pushpa A, Prabha H. A review on xylooligosaccharides. IRJP. 2012;3:71-4.

19. Ebringerova A, Hromadkova Z. Xylans of industrial and biomedical importance, In: Biotechnology and Genetic Engineering Reviews, Intercept Ltd Scientific, Technical & Medical Publishers, Andover. 1999;325-46. ISBN 0264-8725.

20. Blagoveschenski AV, Yurgenson MP. On the changes of wheat proteins under the action of flour and yeast enzymes. Scientific Research Institute of Bread Making, Moscow, U.S.S.R. Biochem J. 1935;29:805-10.

21. Shah AR, Shah RK, Madamwar D. Improvement of the quality of whole wheat bread by supplementation of xylanase from *Aspergillus foetidus*. Bioresour Technol. 2006;97:2047-53.

22. Jiang Z, Li X, Yang S, Li L, Tan S. Improvement of the breadmaking quality of wheat flour by the hyperthermophilic xylanase B from *Thermotoga maritime*. Food Res Int. 2005;38:37-43.

23. Kirk TK, Yang HH. Partial delignification of unbleached kraft pulp with ligninolytic fungi. Biotechnol Lett. 1979;1:347-352.

24. Paice MG, Jurasek L. Removing hemicellulose from pulps by specific enzymic hydrolysis. J Wood Chem Technol. 1984;4:187-98.

25. Converti A, Perego P, Dominguez JM. Xylitol production from hardwood hemicelluloses hydrolysates by *Pachysolen tannophilus, Debaryomyces hansenii* and *Candida guilliermondii*. Appl Biochem Biotechnol. 1999;82:141-51.

26. Chen X, Jiang ZH, Chen S, Qin W. Microbial and bioconversion production of D-xylitol and its detection and application. Int J Biol Sci. 2010;6:834–44.

27. Bailey P. Microbial xylanolytic systems. Trends Biotechnol. 1985;3:286-90.

28. Wong KKY, Saddler JN. *Trichoderma* xylanases, their properties and application. Crit Rev Biotechnol. 1992;12:413-35.

29. Fu JJ, Yang XX, Yu CW.Preliminary research on bamboo degumming with xylanase. Biocatal Biotransform. 2008;25: 450-4.

30. Huang J, Wang G, Xiao L. Cloning, sequencing and expression of the xylanase gene from a *Bacillus subtilis* strain B10 in *Escherichia coli*. Bioresour Technol. 2006;97:802-8.

31. Manenoi A, Paull RE. Papaya fruit softening, endoxylanase gene expression, protein and activity. Physiol Plant. 2007;13:470-80.

32. Guo G, Liu Z, Xu J, Liu J, Dai X, Xie D, etc.Purification and characterization of a xylanase from *Bacillus subtilis* isolated from the degumming line. J Basic Microbiol.2012;52:419-28.

33. Bai W, Xue Y, Zhou C, Ma Y.Cloning, expression and characterization of a novel alkali-tolerant xylanase from alkaliphilic *Bacillus* sp. SN5. Biotechnol Appl Biochem. 2014. DOI: 10.1002/bab.1265.

34. Goswami GK, Pathak RR, Krishnamohan M, Ramesh B. Production, partial purification and biochemical characterization of thermostable xylanase from *Bacillus brevis*. Biomed Pharmacol J. 2013;6:435-40.

35. Dheeran P, Nandhgopal N, Kumar S, Jaiswal YK, Adhikari DK. A novel thermostable xylanases of *Paenibacillus macerans* IIPSP3 isolated from termite gut. J Ind Microbiol Biotechnol. 2012;39:851-60.

36. Lin XQ, Han SY, Zhang N, Hu H, Zheng SP, Ye YR, etc.Bleach boosting effect of xylanase A from *Bacillus halodurans* C-125 in ECF bleaching of wheat straw pulp. Enzyme Microb Technol.2013;52:91-8.

37. Zheng HC, Sun MZ, Meng LC, Pei HS, Zhang XQ, Yan Z, etc. Purification and characterization of a thermostable xylanase from *Paenibacillus* sp. NF1 and its application in xylooligosaccharides production. J Microbiol Biotechnol. 2014; 24:489-96.

38. Valenzuela SV, Diaz P, Pastor FIJ. Xyn11E from *Paenibacillus barcinonensis* BP-23: A LppX-chaperone-dependent xylanases with potential for upgrading paper pulps. Appl Microbiol Biotechnol. 2014;98:5949-57.

39. ul Haq I, Hussain Z, Khan MA, Muneer B, Afzal S, Majeed S, etc. Kinetic and thermodynamic study of cloned thermostable endo-1,4-β-xylanase from *Thermotoga petrophila* in mesophilic host. Mol Biol Rep. 2012;39:7251-61.

40. Zheng H, Liu Y, Liu X, Wang J, Han Y, Lu F. Isolation, purification and characterization of thermostable xylanase from novel strain, *Paenibacillus campinasensis* G1-1. J Microbiol Biotechnol. 2012;22:930-8.

41. Matteotti C, Bauwens J, Brasseur C, Tarayre C, Thonart P, Destain J, etc.Identification and characterization of a new xylanase from Gram-positive bacteria isolated from termite gut (*Reticulitermes santonensis*). Protein Expr Purif. 2012; 83:117-27.

42. Chen YC, Chiang YC, Hsu FY, Tsai LC, Cheng HL. Structural modelling and further improvement in pH stability and activity of a highly-active xylanase from an uncultured rumen fungus. Bioresour Technol. 2012;123:125-34.

43. Tian B, Xu Y, Cai W, Huang Q, Gao Y, Li X, etc. Molecular cloning and overexpression of an endo-β-1,4-xylanase gene from *Aspergillus niger* in industrial *Saccharomyces cerevisiae* YS2 strain. Appl Biochem Biotechnol. 2013;170:320-8.

44. Fan G, Katrolia P, Jia H, Yang S, Yan Q, Jiang Z. High-level expression of a xylanase gene from the thermophilic fungus *Paecilomyces thermophila* in *Pichia pastoris*. Biotechnol Lett. 2012;34:2043-8.

45. Zheng J, Guo N, Wu L, Tian J, Zhou H. Characterization and constitutive expression of a novel endo 1,4- β-d-xylanohydrolase from *Aspergillus niger* in *pichia pastoris*. Biotechnol Lett. 2013;35: 1433-40.

46. Zhang H, Li J, Wang J, Yang Y, Wu M. Determinants for the improved thermostability of a mesophilic family 11 xylanase predicted by computational methods. Biotechnol Biofuels. 2014;7:3.

47. Li J, Wang J, Wang S, Xing M, Yu S, Liu G. Achieving efficient protein expression in *Trichoderma reesei* by using strong constitutive promoters. Microb Cell Fact. 2012;11:84.

48. Song Y, Lee YG, Choi IS, Lee KH, Cho EJ, Bae HJ.Heterologous expression of endo-1,4-β-xylanase A from *Schizophyllum commune* in *Pichia pastoris* and functional characterization of the recombinant enzyme. Enzyme Microb Technol. 2013;52:170-6.

49. Driss D, Bhiri F, Ghorbel R, Chaabouni SE. Cloning and constitutive expression of His-tagged xylanase GH 11 from *Penicillium occitanis* Pol6 in *Pichia pastoris* X33 purification and characterization. Protein Expr Purif. 2012;83:8-14.

50. Xiao Z, Grosse S, Bergeron H, Lau PCK. Cloning and characterization of first GH10 and GH11 xylanases from *Rhizopus oryzae*. Appl Microbiol Biotechnol; 2014. DOI: 10.1007/s-00253-014-5741-4

51. Li XR, Xu H, Xie J, Yi QF, Li W, Qiao DR, etc. Thermostable sites and catalytic characterization of xylanase XYNB of *Aspergillus niger* SCTCC 400264. J Microbiol Biotechnol. 2014;24:483-8.

52. Weerachavangkul C, Laothanachareon T, Boonyapakron K, Wongwilaiwalin S, Nimchua T, Eurwilaichitr L, etc.Alkaliphilic endoxylanase from lignocellulolytic microbial consortium metagenome for biobleaching of eucalyptus pulp. J Microbiol Biotechnol. 2012;22:1636-43.

53. Subramaniyan S. Isolation, purification and characterization of low molecular weight xylanase from *Bacillus pumilus* SSP-34. Appl Biochem Biotechnol. 2012;166:1831-42.

54. Poosarla VG, Chandra TS.Purification and characterization of novel halo-acid-alkali-thermo-stable xylanase from *Gracilibacillus* sp. TSCPVG. Appl Biochem Biotechnol. 2014;173:1375-90.

55. Song L, Dumon C, Siguier B, Andre I, Eneyskaya E, Kulminskaya A, etc. Impact of an N-terminal extension on the stability and activity of the GH11 xylanase from *Thermobacillus xylanilyticus*. J Biotechnol. 2014;174:64-72.

56. Wang J, Sun Z, Zhou Y, Wang Q, Ye J, Chen Z, etc. Screening of a xylanase clone from a fosmid library of rumen microbiota in Hu sheep. Anim Biotechnol. 2012;23:156-73.

57. Shan ZQ, Zhou JG, Zhou YF, Yuan HY, Hong LV.Isolation and characterization of an alkaline xylanases from a newly isolated *Bacillus* sp. QH14. Hereditas (Beijing). 2012;34:356-65.

58. Kamble RD, Jadhav AR. Production, purification and characterization of alkali stable xylanase from *Cellulosimicrobium* sp, MTCC 10645. Asian Pac J Trop Biomed. 2012;2:1790-7.

59. Wang K, Luo H, Tian J, Turunen O, Huang H, Shi P, etc. Thermostability improvement of a *streptomyces* xylanase by introducing proline and glutamic acid residues. Appl Environ Microbiol. 2014;80:2158-65.

60. Kim DY, Ham SJ, Kim HJ, Kim J, Lee MH, Cho HY, etc. Novel modular endo-β-1,4-xylanase with transglycosylation activity from *Cellulosimicrobium* sp. strain HY-13 that is homologous to inverting GH family 6 enzymes. Bioresour Technol. 2012;107:25-32.

61. Baek CU, Lee SG, Chung YR, Cho I, Kim JH. Cloning of a family 11 xylanase gene from *Bacillus amyloliquefaciens* CH51 isolated from Cheonggukjang. Indian J Microbiol. 2012;52:695-700.

62. Singh RK, Tiwari MK, Kim D, Kang YC, Ramachandran P, Lee JK. Molecular cloning and characterization of GH11 endoxylanase from *Chaetomium globosum*, and its use in enzymatic pretreatment of biomass. Appl Microbiol Biotechnol. 2013;97:7205-14.

63. Jagtap SS, Dhiman SS, Kim TS, Li J, Lee JK, Kang YC. Enzymatic hydrolysis of aspen biomass into fermentable sugars by using lignocellulases from *Armillaria gemina*. Bioresour Technol. 2013;133:307-14.

64. Dhiman SS, Kalyani D, Jagtap SS, Haw JR, Kang YC, Lee JK. Characterization of a novel xylanase from *Armillaria gemina* and its immobilization onto SiO2 nanoparticles. Appl Microbiol Biotechnol. 2012;97:1081-91.

65. Wang J-Q, Yin X, Wu M-C, Zhang H-M, Gao S-J, Wei J-T, etc.Expression of family 10 xylanase gene from *Aspergillus usamii* E001 in *Pichia pastoris* and characterization of the recombinant enzyme. J Ind Microbiol Biotechnol. 2013;40:75-83.

66. Zhang HM, Wang JQ, Wu MC, Gao SJ, Li JF, Yang YJ. Optimized expression, purification and characterization of a family 11 xylanase (AuXyn11A) from *Aspergillus*

usamii E001 in *Pichia pastoris*. J Sci Food Agric. 2013;94:699-706.

67. Do TT, Quyen DT, Nguyen TN, Nguyen VT.Molecular characterization of a glycosyl hydrolase family 10 xylanase from *Aspergillus niger*. Protein Expr Purif. 2013;92:196-202.

68. Kim HM, Lee KH, Kim KH, Lee DS, Nguyen QA, Bae HJ.Efficient function and characterization of GH10 xylanase (Xyl10g) from *Gloeophyllum trabeum* in lignocellulose degradation. J Biotechnol. 2014;172:38-45.

69. Available:http://bch.cbd.int/protocol/backgr ound/ (Accessed on 26 August 2014).

70. Available:http://bch.cbd.int/protocol/parties/ (Accessed on 26 August 2014).

71. Secretariat of the Convention on Biological Diversity. Cartagena Protocol on Biosafety to the Convention on Biological Diversity: text and annexes; 2000. Available:http://bch.cbd.int/database/attach ment/?id=10694 (Accessed on 26 August 2014).

72. Secretariat of the Convention on Biological Diversity. Nagoya - Kuala Lumpur Supplemetary Protocol on Liability and Redress to the Cartagena Protocol on Biosafety; 2011. Available:http://bch.cbd.int/database/attach ment/?id=11064 (Accessed on 26 August 2014).

73. Presidency of the Republic Brazil Sub-Office of Legal Affairs. Law No. 11.105,2005. Available:http://www.wipo.int/edocs/lexdoc s/laws/en/br/br060en.pdf (Accessed on 26 August 2014).

74. WTO Committee on Technical Barriers to Trade. Notification on Ministerial Act "Portaria No. 2.658"; 2004. Available:http://www.inmetro.gov.br/barreir astecnicas/pontofocal/textos/notificacoes/B RA_119_add_1.pdf (Accessed on 26 August 2014).

75. Ministry of Agriculture of the People's Republic of China. Regulations on Administration of Agricultural Genetically Modified Organisms Safety; 2011. Available:http://english.agri.gov.cn/hottopic s/bt/201301/t20130115_9551.htm (Accessed on 26 August 2014).

76. Wang M. Legal Issues Relating to GMO Safety in China. Environmental Law Institute; 2007. Available:http://elr.info/sites/default/files/art icles/37.10817.pdf (Accessed on 26 August 2014).

77. European Parliament and Council. Consolidated version of the Treaty on the Functioning of the European Union; 2012. Available:http://eur-lex.europa.eu/legal-content/EN/TXT/HTML/?uri=CELEX:12012 E/TXT&from=DE (Accessed on 26 August 2014).

78. European Parliament and Council. Regulation (EC) No 1907/2006 of the European Parliament and of the Council; 2006. Available:http://eur-lex.europa.eu/LexUriServ/LexUriServ.do?u ri=OJ:L:2007:136:0003:0280:en:PDF (Accessed on 26 August 2014).

79. European Parliament and Council. Directive 2009/41/EC of the European Parliament and of the Council; 2009. Available:http://eur-lex.europa.eu/legal-content/EN/TXT/PDF/?uri=CELEX:32009L 0041&from=en (Accessed on 26 August 2014).

80. European Parliament and Council. Directive 2001/18/EC of the European Parliament and of the Council; 2001. Available:http://eur-lex.europa.eu/resource.html?uri=cellar:303 dd4fa-07a8-4d20-86a8-0baaf0518d22.0004.02/DOC_1&format=P DF (Accessed on 26 August 2014).

81. European Parliament and Council. Regulation (EC) No 1829/2003 of the European Parliament and of the Council; 2003. Available:http://eur-lex.europa.eu/legal-content/en/TXT/HTML/?uri=CELEX:32003 R1829&from=en (Accessed on 26 August 2014).

82. Fish A, Rudenko L. Guide to U.S. Regulation of Genetically Modified Food and Agricultural Biotechnology Products. The Pew Charitable Trusts; 2001. Available:http://www.pewtrusts.org/~/medi a/legacy/uploadedfiles/wwwpewtrustsorg/r eports/food_and_biotechnology/hhsbiotech 0901pdf.pdf (Accessed on 26 August 2014).

83. Environmental Protection Agency Title 40 Chapter I Subchapter R part 725 - Reproting Requiremnets and Review Processesfor Microorganisms. Available:http://www.ecfr.gov/cgi-bin/text-idx?SID=e33fe8c4ee10e689c37b97fc05dc

d393&node=40:31.0.1.1.13&rgn=div5#40:
31.0.1.1.13.3.1.3. (Accessed on 26 August
2014).

84.   US EPA TSCA Chemical Substance
      Inventory.
      Available:http://www.epa.gov/oppt/existing
      chemicals/pubs/tscainventory/howto.html.
      (Accessed on 26 August 2014).

85.   Rulis A. Agency Response Letter GRAS
      Notice No. GRN 000054. Environmental
      Protection   Agency;   2001.
      Available:http://www.fda.gov/Food/Ingredie
      ntsPackagingLabeling/GRAS/NoticeInvent
      ory/ucm153739.htm (Accessed on 26
      August 2014).

16

Identification of a Biochemical Indicator in Plants for Heavy Metal Contamination

Sonali Hazra Das[1], Mahasweta Paul[1], Shankha Ghosh Dastidar[1], Tulika Das[1] and Srabanti Basu[1*]

[1]Department of Biotechnology, Heritage Institute of Technology, Kolkata, India.

Authors' contributions

The work has been carried out in collaborations between all authors. Authors SHD, MP, SGD and TD managed literature survey, performed the experiments and did the statistical analysis. Authors MP and SGD wrote the first draft of the manuscript. Author SB designed the experiments, supervised the study and prepared the final manuscript. All authors read and approved the final manuscript.

Editor(s):
(1) Juan Pedro Navarro – Aviñó , Technical University of Valencia, Spain.
(2) Kuo-Kau Lee, Department of Aquaculture, National Taiwan Ocean University, Taiwan.
Reviewers:
(1) Anonymous, Ireland.
(2) Echem, Ogbonda G, Science Laboratory Technology, Rivers State Poly Technic, Bori, Nigeria.
(3) F.B.G.Tanee, Plant Science and Biotechnology Department, University of Port Harcourt, Nigeria.

ABSTRACT

Aims: To identify a biochemical parameter in a plant(s) that can serve as an indicator of heavy metal pollution.
Study Design: Design of the study includes collection of plant samples from a contaminated site and a rural site and to check their biochemical parameters related to oxidative stress. Based on the experimental findings, attempts were made to identify a parameter that can serve as an indicator of heavy metal contamination.
Place and Duration of Study: The study was conducted in Department of Biotechnology, Heritage Institute of Technology, India, during the period from August 2013-June 2014.
Methodology: Plant samples were collected from East Calcutta Wetlands (metal contaminated site) and Binogram, a village 90 Km away from Kolkata (control site). The samples were checked for the following parameters: non-protein thiol, ascorbic acid, superoxide dismutase (SOD), lipid peroxidation. This was followed by a statistical analysis.
Results: Non-protein thiol content was found to be less in the *Ipomoea, Hygrophila* and cabbage

---

samples collected from ECW. The values were 65-75% of the village samples. Cauliflower leaves did not show any alteration in the parameter. Ascorbic acid was altered significantly in cabbage and *Hygrophila*. Activity of SOD decreased in all samples. Lipid peroxidation increased in all samples except *Ipomoea*. Of the biochemical parameters tested, maximum alteration was found with non-protein thiol, in which glutathione (GSH) is the major constituent. Of the vegetables studied, cabbage showed an alteration in all biochemical parameters.

Conclusion: GSH has the potential for being used as an indicator of heavy metal pollution. Cabbage can be used as a model vegetable for testing the biochemical parameters. The method will be helpful in identification of a contaminated site before estimation of heavy metals.

Keywords: Oxidative stress; non-protein thiol; East Calcutta Wetlands; heavy metals; plants.

1. INTRODUCTION

Agricultural soils in many parts of the world are slightly-to-moderately contaminated by heavy metals. This could be due to the long term use of phosphatic fertilizers, sewage sludge applications, dust from smelters, industrial waste and bad watering practices in agricultural lands [1-3]. With industrial development, the production and emission of heavy metals have increased. Some metals like arsenic, cadmium, chromium, lead etc., when present at high concentrations, have strong toxic effects and pose an environmental threat. Presence of heavy metals in different food causes serious health hazards, depending on their relative levels.

The primary response of plants toward stress is the generation of Reactive Oxygen Species (ROS) upon exposure to high levels of heavy metals. Various metals either generate ROS directly through Haber-Weiss reactions or create oxidative stress through indirect mechanisms [4,5]. The indirect mechanisms include their interaction with the antioxidant system, disrupting the electron transport chain or disturbing the metabolism of essential elements [6-8]. To countermine the toxicity of ROS, living organisms possess defence systems called anti-oxidative or anti-oxidant defence systems, comprising both enzymatic and non-enzymatic constituents. The non-enzymatic anti-oxidants are generally small molecules that include the tripeptide glutathione, cysteine, hydroxy-quinone, ascorbate, the lipophilic anti-oxidants α-tocopherol, carotenoid pigments and a variety of other compounds [9]. The enzymatic anti-oxidant components include the enzymes capable of removing, neutralizing or scavenging ROS and the intermediates, such as catalase, glutathione peroxidase, ascorbate peroxidase, superoxide dismutase, glutathione reductase, monodehydroascorbate reductase and dehydroascorbate reductase. The non-enzymatic and enzymatic components work in close co-ordination for the effective removal of ROS. Thus, a balance is maintained between the formation and destruction of ROS. A shift in the balance between pro-oxidative and anti-oxidative systems will lead to Oxidative Stress or accumulation of the toxic ROS.

East Calcutta Wetlands (ECW), located approximately between latitudes 22°25' to 22°40' N to longitudes 88°20' to 88°35' E covering 12,500 hectares area, acts as the sewage treatment and resource recovery system of Kolkata. The resource recovery takes place mainly by three processes – sewage-fed fisheries, agriculture on the garbage substrate and paddy cultivation using pond effluent. The Kolkata Municipality generates 600 million litres of wastewater daily. The wastewater flows through underground sewers to the pumping stations at the eastern fringes of the city and is then pumped into open channels. The wastewater is received by the sewage-fed fisheries and local farms. The wetland has been incorporated in the list maintained by the Ramsar Bureau established under the Ramsar Convention and is recognised as a 'Wetland of International Importance'. The site receives mixed waste containing domestic wastewater as well as industrial effluents. Reports are available on presence of heavy metals in the soil of this area [10,11].

The objective of the present study is to check the biochemical stress parameters of plants grown in the East Calcutta Wetlands and to identify a biochemical parameter that can serve as an indicator of heavy-metal contamination.

2. MATERIALS AND METHODS

2.1 Collection of Samples

Plant and soil samples were collected from the East Calcutta Wetlands (experimental site or the site contaminated with heavy metals), and Binogram, a village of Hooghly district, 90 Km away from Kolkata (control site). Location map of the sample collection site is given in Fig. 1. Plant samples included *Hygrophila auriculata*, *Ipomoea aquatica*, *Brassica oleracea* (Cauliflower: Botrytis cultivar group), *Brassica oleracea* (Cabbage: White cabbage: *B. oleracea* L. *var. capitata* L. *f. alba DC.*). The plant samples were collected in the season of their cultivation and the biochemical tests were carried out within five days of collection. Nine samples of each plant were tested for each biochemical parameter.

2.2 Estimation of Heavy Metals in Soil Samples

Soil samples were dried keeping it at 70ºC and weighed periodically till constant weight is achieved. The dried samples were crushed and digested with concentrated hydrochloric acid and 70% perchloric acid. Arsenic, chromium, cadmium and lead were estimated in the digests by atomic absorption spectrophotometry.

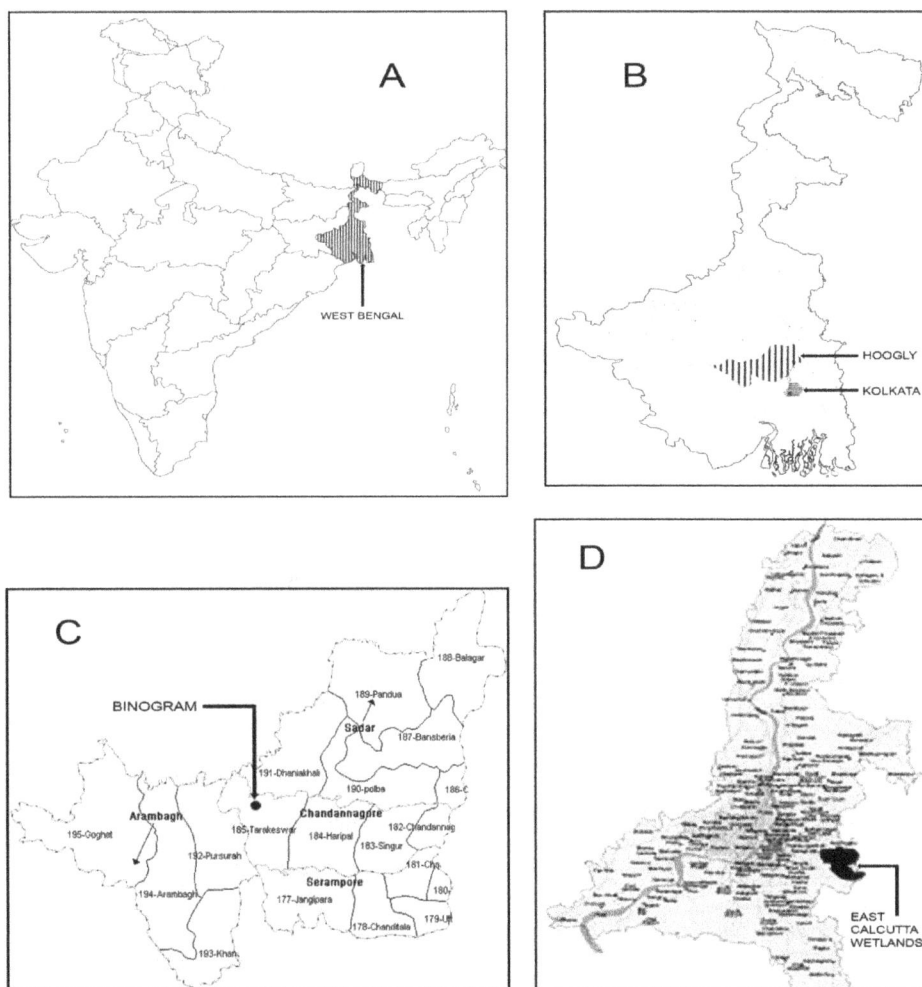

Fig. 1. Location map of the sample collection site (A): location of West Bengal in India, (B): location of Hooghly district and Kolkata in West Bengal, (C) location of Binogram in Hooghly, (D): location of East Calcutta Wetlands in Kolkata

2.3 Estimation of Non-protein Thiol

Plant extract (20% w/v) was prepared in 6% (w/v) TCA and centrifuged at 5,000 rpm for 10 minutes to separate the proteins. Non-protein thiol was estimated in the supernatant spectrophotometrically (Shimadzu UV1700) using 5,5'di-thio-bis [nitro-benzoic acid] (DTNB) reagent [12]. Absorbance was measured at 440 nm (Shimadzu UV1700).

2.4 Estimation of Ascorbic Acid

Aqueous extracts of plant samples (30% w/v) were centrifuged at 5000 rpm for 5 minutes and the supernatant was collected. Ascorbic acid was estimated by volumetric method following the procedure of Madhava-Rao [13] using 2, 6 dichlorophenol indophenol as an indicator.

2.5 Assay of Superoxide Dismutase (SOD)

Aqueous extracts of plant samples (20% w/v) were centrifuged at 5,000 rpm for 5 minutes and the supernatant was collected. SOD was estimated spectrophotometrically following the prevention of auto-oxidation of pyrogallol [14]. A reduction of 50% auto oxidation of pyrogallol is considered 1 unit of SOD.

2.6 Estimation of Lipid Peroxidation

Plant extract (20% w/v) was made in 0.5M saline-phosphate buffer, pH 7.4. Lipid peroxidation was estimated using thiobarbituric acid (TBA). Determination of TBA-reactive substances (TBARS), mostly consisting of malonaldehyde, was performed spectrophotometrically with the supernatant [15].

Statistical analysis of all the experimental data was done using Student's t test (unpaired).

3. RESULTS AND DISCUSSION

3.1 Estimation of Heavy Metals in Soil Samples

Soil samples collected from the East Calcutta Wetlands are highly contaminated with heavy metals as found from the study. Soil samples of the control site contained much lower amounts of chromium, cadmium and arsenic than the experimental site (Table 1). However, a considerable amount of lead is present in the soil sample of the control site which might have come from exhausts of automobiles passing through the village roads. However, the amount is lower than that found in the soil sample of the experimental site.

3.2 Estimation of Non-protein Thiol

The non-protein thiol content was decreased in *Ipomoea*, *Hygrophila* and cabbage samples collected from the ECW (Fig. 2). The decrease was comparable in cabbage and *Hygrophila* (74% and 75% respectively of the control-site), whereas decrease in *Ipomoea* was a little less (65% of the control-site). Cauliflower was an exception where no change in thiol content was found. The finding is supported by Schutzendubel [16] who reported decrease in glutathione (GSH), the major non-protein thiol present in living systems, due to cadmium stress.

GSH is a key component in metal scavenging mechanisms due to its high affinity towards heavy metals, especially arsenic, lead, cadmium and mercury. Apart from binding the heavy metals directly, GSH serves as the precursor of phytochelatin, a protein synthesized under the condition of heavy-metal stress [17]. Induction of phytochelatin synthesis has been best reported in cases of cadmium toxicity. However, it can be induced in the presence of other heavy metals like mercury, copper, zinc, lead and nickel [18]. The depletion of non-protein thiols in the plant samples collected from ECW may have resulted from higher synthesis of phytochelatin in those plants due to the presence of heavy metals in soil.

3.3 Estimation of Ascorbic Acid

Ascorbic acid has been decreased significantly in cabbage and *Hygrophila* collected from the ECW and remained unchanged in cauliflower and *Ipomoea* (Fig. 3). The decrease was comparable in both the samples (10% and 12% of the control sites respectively). Ascorbic acid is a component of anti-oxidant defence mechanism and plays an important role in protection of plants against several environmental stresses including heavy metal stress [19].

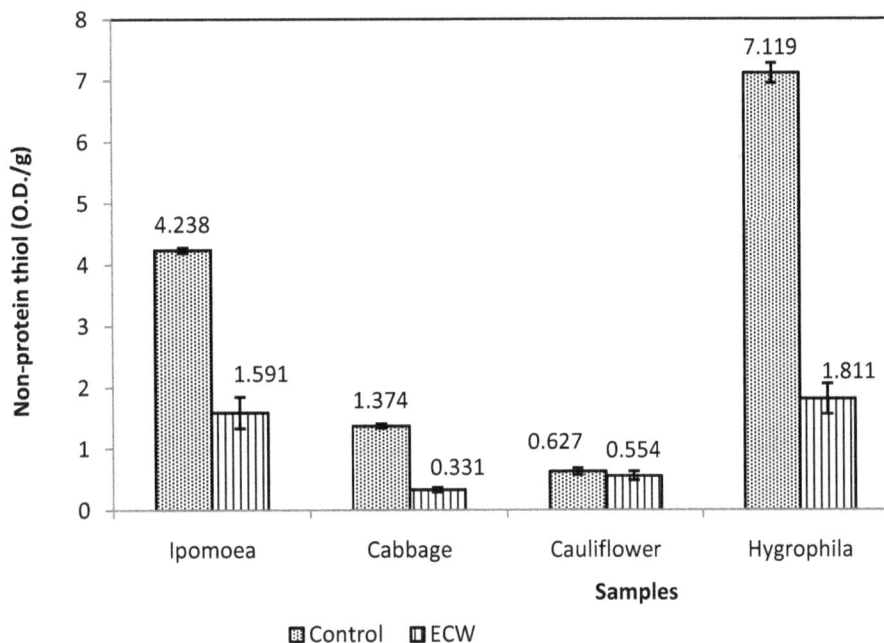

Fig. 2. Non-protein thiol-contents in plant samples collected from East Calcutta Wetlands and the control sites

Significant changes: Ipomoea (P<0.0001) cabbage (P<0.0001), hygrophila (P<0.001)

Table 1. Heavy metal content of the soil samples

Metal	Binogram (control site) (mg/kg)	East Calcutta Wetlands (experimental site) (mg/kg)
Chromium	6.935	31.23
Cadmium	1.395	2.16
Lead	30.9	60.3
Arsenic	<0.2	0.79

Previous studies have shown that ascorbic acid content in the roots and shoots of two cultivars of pigeon pea (*Cajanus cajan*) decreased when zinc and nickel was supplied from outside. Moreover, ascorbic acid content showed a significant negative correlation with increasing concentrations of the two metal ions [13]. Soto et al. [20] have reported lower ascorbic acid level in *M. muscorum* in presence of organic compounds. Reduced ascorbic acid content may be due to direct oxidation of ascorbic acid caused by lead (II). It is also possible that ascorbic acid has been consumed by the antioxidant defence enzyme ascorbate peroxidase. The possibility of consumption of ascorbic acid by ascorbate peroxidase can be substantiated by the observation of Okamoto et al. [21]. The group has reported high activity of ascorbate peroxidase content in *Gonyaulax polyedra* due to heavy metal stress induced by lead (II), mercury (II), cadmium (II) and copper (II).

3.4 Assay of SOD

The activity of SOD has been found to be decreased significantly in all three samples tested for it (Fig. 4). In an earlier study, John et al. [22] have reported higher SOD activity due to exposure to cadmium at both low and high concentrations. However, Zhang et al. [23], who studied heavy metal stress on oxidative enzymes in leaves of two mangrove plant seedlings, namely *Kandelia candel* and *Bruguiera gymnorrhiza*, reported that SOD activity fluctuated in plants under different stressed conditions. From the above information from other studies and the observation of the present study, it can be stated that SOD may not be a good parameter to be used as biochemical indicator of heavy metal stress in plants.

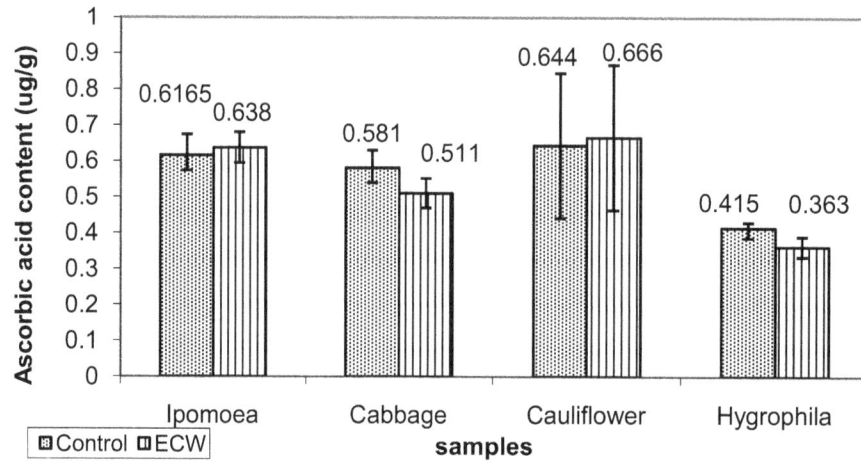

Fig. 3. Ascorbic acid contents in plant samples collected from East Calcutta Wetlands and the control sites
Significant changes: cabbage (P<0.005), hygrophila (P<0.05)

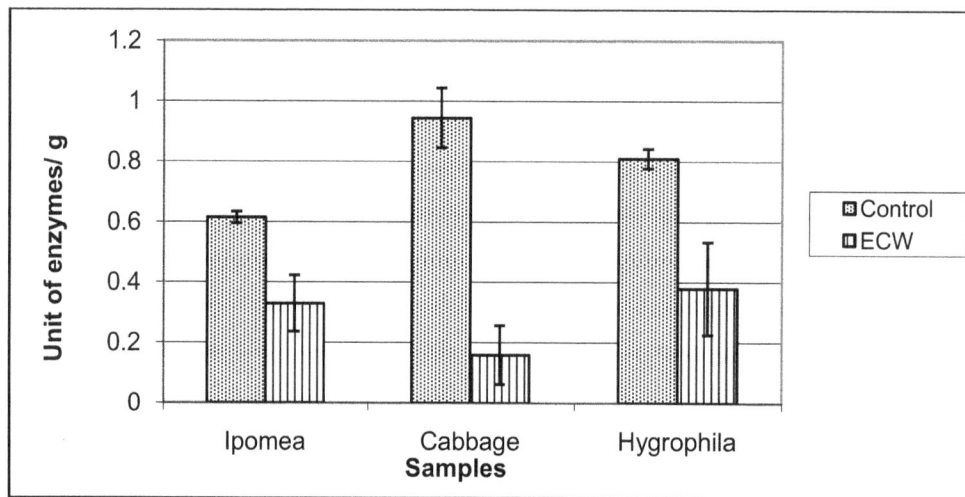

Fig. 4. Activity of superoxide dismutase (SOD) in plant samples collected from East Calcutta Wetlands and the control sites
Significant changes: Ipomoea (p<0.01), cabbage (p<0.005), hygrophila (P<0.005)

3.5 Estimation of Lipid Peroxidation

Lipid peroxidation was significantly increased in cabbage, cauliflower and *Hygrophila* and remained unaltered in *Ipomoea* (Fig. 5). There is a possibility of enhanced lipid peroxidation in tissues as heavy metals are known to induce oxidative stress in plants. The reason may be that heavy metals, particularly the transition metals, can induce production of hydroxyl radicals which could lead to lipid peroxidation [16].

High lipid peroxidation was observed by Zhang et al. [23] in the leaves of *Bruguiera gymnorrhiza* under heavy metal stress. Excess cobalt, nickel and cadmium produced increased amount of TBARS in spinach [24]. John et al. [22] have reported high amount of malondialdehyde, a product of lipid peroxidation, in *Lemna polyrrhiza* due to cadmium stress. The increase in lipid peroxidation may be a direct effect of the metal toxicity caused by increased production of ROS or an indirect effect caused by depletion of GSH and/or components of anti-oxidant defence mechanism.

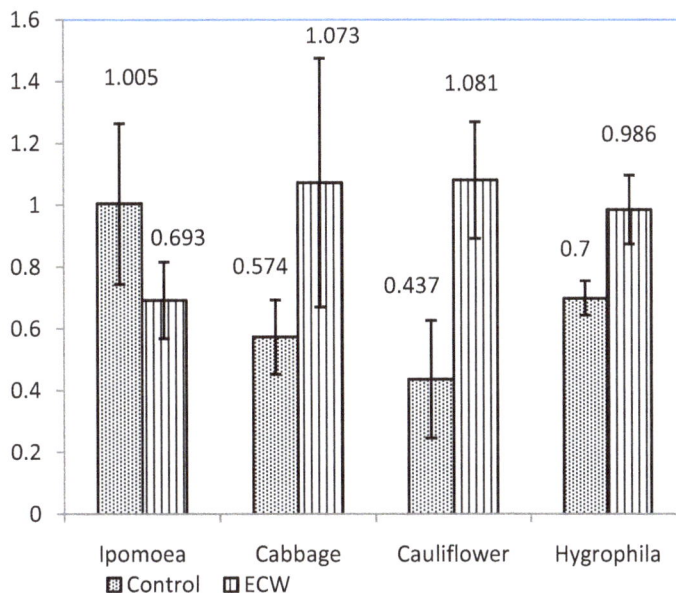

Fig. 5. Lipid peroxidation in plant samples collected from East Calcutta Wetlands and the control sites
Significant changes: cabbage (P<0.01), cauliflower (P<0.0002), hygrophila (P<0.002)

4. CONCLUSION

Of the biochemical parameters tested, maximum alteration was found with the parameter non-protein thiol, in which GSH is the major constituent. Therefore it can be concluded that GSH content of plant samples has the potential for being used as an indicator of heavy metal pollution of a contaminated site. Of the vegetable studied, cabbage showed an alteration in all biochemical parameters. Hence cabbage perhaps can be used as a model plant species in which the number of parameters showing variability is higher in a contaminated site among the species studied; thereby indicating a maximum level of sensitivity in regard to those parameters. However, more field studies are required to validate the present observation. Moreover, GSH content and other biochemical parameters are required to be tested under laboratory conditions. The process will be helpful for identification of a contaminated site before going for estimation and identification of heavy metals.

COMPETING INTERESTS

Authors have declared that no competing interests exist.

REFERENCES

1. Bell FG, Bullock SET, Halbich TFJ, Lindsay P. Environmental impacts associated with an abandoned mine in the Witbank Coalfield, South Africa. Int J Coal Geol. 2001;45:195–216.

2. Schwartz C, Gerard E, Perronnet K, Morel JL. Measurement of in-situ phytoextraction of zinc by spontaneous metallophytes growing on a former smelter site. Sci Total Environ. 2001;279:215–21.

3. Passariello B, Giuliano V, Quaresima S, Barbaro M, Caroli S, Forte G Garelli G Iavicoli I. Evaluation of the environmental contamination at an abandoned mining site. Microchem J. 2001;73:245–50.

4. Wojtaszek P. Oxidative burst: An early plant response to pathogen infection. Biochem J. 1997;322:681–92.

5. Mithofer A, Schulze B, Boland W. Biotic and heavy metal stress response in plants: Evidence for common signals. FEBS Lett. 2004;566:1–5.

6. Srivastava S, Tripathi RD, Dwivedi UN. Synthesis of phytochelatins and modulation of antioxidants in response to cadmium stress in Cuscuta reflexa—an angiospermic parasite. J Plant Physiol. 2004;161:665–74.

7. Qadir S, Qureshi MI, Javed S, Abdin MZ. Genotypic variation in phytoremediation potential of Brassica juncea cultivars exposed to Cd stress. Plant Sci. 2004;167:1171–81.

8. Dong J, Wu FB, Zhang GP. Influence of cadmium on antioxidant capacity and four microelement concentrations in tomato seedlings (*Lycopersicon esculentum*). Chemosphere. 2006;64:1659–66.

9. Larson RA. The antioxidants of higher plants. Phytochemistry. 1988;27:969-78.

10. Raychaudhuri S, Mishra M, Nandy P, Thakur A R. Waste management: A case study of ongoing traditional practices in East Calcutta Westland. Am J Agric Biol Sci. 2008;3(1):315-20.

11. Kundu N, Pal M, Saha S. East Kolkata Wetlands: A resource recovery system through productive activities. Proceedings of Taal 2007: The 12th World Lake Conference. 2008;868-81.

12. Owens CWI, Belcher RVA. Colorimetric micro-method for the determination of glutathione. Biochem J. 1965;94:705–11.

13. Madhava Rao, Sresty TVS. Antioxidative Parameters in the seedlings of Pigeonpea (*Cajanus cajan* (L.) Millspaugh) in response to Zn and Ni stresses. Plant Sci. 2000;1572:113-28

14. Marklund S, Marklund G. Involvement of superoxide anion radical in the auto-oxidation of pyrogallol and a convenient assay for superoxide. Eur J Biochem. 1972;47:3170-5.

15. Janero D. Malonaldehyde and thiobarbituric acid reactivity as diagnostic indices of lipid peroxidation and peroxidative tissue injury. Free Rad Biol Med. 1990;9:515-40.

16. Schutzendubel A, Polle A. Plant responses to abiotic stresses: Heavy metal-induced oxidative stress and protection by mycorrhization. J Exp Bot. 2002;53:1351-65.

17. Jozefezak M, Remans T, Vangronsveld J, Cuypers A. Glutathione is a key player in metal-induced oxidative stress defences. Int J Mol Sc. 2012;13:3145-75.

18. Yadav SK. Heavy metals toxicity in plants: An overview on the role of glutathione and phytochelatins in heavy metal stress tolerance of plants. South African J Bot. 2010;76:167-79.

19. Khan TA, Mazid M, Mohammad F. A review of ascorbic acid potentialities against oxidative stress induced in plants. J Agrobiol. 2011;28(2):97-111.

20. Soto P, Gaetel G, Hidalgo ME. Assessment of catalase activity, lipid peroxidation, chlorophyll-a, and growth rate in the fresh water green algae *Pseudokirchneriella subcapitala* exposed to copper and zinc. Lat Am Aquat Res. 2011;39(2):280-5.

21. Okamoto M, Kuwahara A, Seo M, Kushiro T, Asami T, Hirai N et al. CYP707A1 and CYP707A2, Which encode abscisic acid 8-hydroxylases, are iIndispensable for proper control of seed dormancy and germination in arabidopsis. Plant Physiol. 2006;141:97-107.

22. John R, Ahmad P, Gadgil K, Sharma S. Antioxidative response of *Lemma polyrrhiza* L. to cadmium stress. J Env Biol. 2007;28(3):583-9.

23. Zhang FQ, Wang YS, Lou ZP, Dong JD. Effect of heavy metal stress on antioxidative enzymes and lipid peroxidation in leaves and roots of two mangrove plant seedlings (Kandeliacandel and Bruguiera gymnorrhiza), Chemosphere. 2007;67:44-50.

24. Pandey N, Pathak G, Pandey D, Pandey R. Heavy metals Co, Ni, Zn and Cd produce oxidative damage and evoke differential antioxidant responses in spinach. Braz J Plant Physiol. 2009;21:103-111. DOI.org/10.1590/S1677-04202009000rg/10.1590/S1677-04202009000.

Permissions

List of Contributors

A. D. Ologhobo
Department of Animal Science, University of Ibadan, Ibadan, Nigeria

I. O. Adejumo
Department of Animal Science, Landmark University, Omu-Aran, P.M.B.1001, Kwara State, Nigeria

Mina Khanbagi Dogahe
Department of Food Science and Technology, Varamin Branch, Islamic Azad University, Varamin, Tehran, Iran

Kianoush Khosravi-Darani
Research Department of Food Technology, National Nutrition and Food Technology Research
Institute, Faculty of Nutrition Sciences and Food Technology, Shahid Beheshti University of Medical Sciences, P.O. Box: 193954741, Tehran, Iran

Azadeh Tofighi
Department of Food Science and Technology, Varamin Branch, Islamic Azad University, Varamin, Tehran, Iran

Mehrnaz Dadgar
Department of Food Science and Technology, Varamin Branch, Islamic Azad University, Varamin, Tehran, Iran

Amir Mohammad Mortazavian
Department of Food Sciences and Technology, National Nutrition and Food Technology Research Institute, Faculty of Nutrition Sciences and Food Technology, Shahid Beheshti University of Medical `Sciences, Tehran, Iran

E. C. Egwim
Biochemistry Department, Global Institute for Bioexploration, Federal University of Technology, P.M.B. 65, Minna, Niger state, Nigeria

T. Caleb
Biochemistry Department, Global Institute for Bioexploration, Federal University of Technology, P.M.B. 65, Minna, Niger state, Nigeria

Karin Vega
Laboratorio de Micologíay Biotecnología, Universidad Nacional Agraria La Molina, Av. La Molina s/n, La Molina, Lima 12, Peru

Victor Sarmiento
Laboratorio de Micologíay Biotecnología, Universidad Nacional Agraria La Molina, Av. La Molina s/n, La Molina, Lima 12, Peru

Yvette Ludeña
Laboratorio de Micologíay Biotecnología, Universidad Nacional Agraria La Molina, Av. La Molina s/n, La Molina, Lima 12, Peru

Nadia Vera
Laboratorio de Micologíay Biotecnología, Universidad Nacional Agraria La Molina, Av. La Molina s/n, La Molina, Lima 12, Peru

Carmen Tamariz-Angeles
Laboratorio de Micologíay Biotecnología, Universidad Nacional Agraria La Molina, Av. La Molina s/n, La Molina, Lima 12, Peru

Gretty K. Villena
Laboratorio de Micologíay Biotecnología, Universidad Nacional Agraria La Molina, Av. La Molina s/n, La Molina, Lima 12, Peru

Marcel Gutiérrez-Correa
Laboratorio de Micologíay Biotecnología, Universidad Nacional Agraria La Molina, Av. La Molina s/n, La Molina, Lima 12, Peru

K. Nyembo
Department of Phytobiology, Division of Life Science, General Atomic Energy Commission, Regional Center of Nuclear Studies of Kinshasa XI, University of Kinshasa, DR, Congo

N. Mbaya
Department of Phytobiology, Division of Life Science, General Atomic Energy Commission, Regional Center of Nuclear Studies of Kinshasa XI, University of Kinshasa, DR, Congo

N. Muambi
Department of Phytobiology, Division of Life Science, General Atomic Energy Commission, Regional Center of Nuclear Studies of Kinshasa XI, University of Kinshasa, DR, Congo

C. Ilunga
Department of Biotechnology, Division of Life Science, General Atomic Energy Commission,
Regional Center of Nuclear Studies of Kinshasa, XI, University of Kinshasa, DR, Congo

Ritu Jain
School of Biotechnology, Devi Ahilya University, Khandwa Road, Indore-452001, India

Sarika Garg
Department of Neuroscience, The University of Montreal Hospital Research Centre (CRCHUM), Bureau/Office: CRCHUM R09.476, Tour Viger, 900, rue St-Denis, Montréal, Québec H2X 0A9, Canada

Anil Kumar
School of Biotechnology, Devi Ahilya University, Khandwa Road, Indore-452001, India

Hailu Weldekiros Hailu
Department of Biology, Bioscience postgraduate program, Faculty of Mathematics and Natural Sciences, Sebelas Maret University, Surakarta, Indonesia
Department of Biology, College of Science, Bahir Dar University, Bahir Dar, Amhara, Ethiopia

Viktor Abdallah
Department of Biology, Bioscience postgraduate program, Faculty of Mathematics and Natural Sciences, Sebelas Maret University, Surakarta, Indonesia

Ari Susilowati
Department of Biology, Bioscience postgraduate program, Faculty of Mathematics and Natural Sciences, Sebelas Maret University, Surakarta, Indonesia

Saleh Muhammed Raqib
Department of Biology, Bioscience postgraduate program, Faculty of Mathematics and Natural Sciences, Sebelas Maret University, Surakarta, Indonesia
Department of Biotechnology, Faculty of Agriculture, Bangladesh Agriculture University, Mymensingh, Bangladesh

Ali Ramadan Ali ALatawi
Department of Chemistry, Faculty of Science, Sirte University, Sirte, Libya

Hailu Weldekiros Hailu
Department of Biology, Bioscience postgraduate program, Faculty of Mathematics and Natural Sciences, Sebelas Maret University, Surakarta, Indonesia
Department of Biology, College of Science, Bahir Dar University, Bahir Dar, Amhara, Ethiopia

Viktor Abdallah
Department of Biology, Bioscience postgraduate program, Faculty of Mathematics and Natural Sciences, Sebelas Maret University, Surakarta, Indonesia

Ari Susilowati
Department of Biology, Bioscience postgraduate program, Faculty of Mathematics and Natural Sciences, Sebelas Maret University, Surakarta, Indonesia

Saleh Muhammed Raqib
Department of Biology, Bioscience postgraduate program, Faculty of Mathematics and Natural Sciences, Sebelas Maret University, Surakarta, Indonesia
Department of Biotechnology, Faculty of Agriculture, Bangladesh Agriculture University, Mymensingh, Bangladesh

Ali Ramadan Ali ALatawi
Department of Biology, Bioscience postgraduate program, Faculty of Mathematics and Natural Sciences, Sebelas Maret University, Surakarta, Indonesia
Department of Chemistry, Faculty of Science, Sirte University, Sirte, Libya

O. R. Ezeigbo
Deparment of Biology/Microbiology, Abia State Polytechnic, Aba, Nigeria

C. F. Ukpabi
Deparment of Chemistry/Biochemistry, Abia State Polytechnic, Aba, Nigeria

C. A. Ike-Amadi
Deparment of Chemistry/Biochemistry, Abia State Polytechnic, Aba, Nigeria

M. U. Ekaiko
Deparment of Chemistry/Biochemistry, Abia State Polytechnic, Aba, Nigeria

H. N. Ogungbenle
Department of Chemistry, Ekiti State University, P.M.B. 5363, Ado - Ekiti, Nigeria

O. T. Oyadipe
Department of Chemistry, Ekiti State University, P.M.B. 5363, Ado - Ekiti, Nigeria

Yves B. Nyamien
Laboratory of Biochemistry and Food Science, Training and Research Unit of Biosciences, Felix HOUPHOUËT-BOIGNY University of Abidjan, 22 BP 582 Abidjan 22, Côte d'Ivoire

Olivier Chatigre
Laboratory of Biochemistry and Food Science, Training and Research Unit of Biosciences, Felix HOUPHOUËT-BOIGNY University of Abidjan, 22 BP 582 Abidjan 22, Côte d'Ivoire

Emmanuel N. Koffi
Laboratory of Bioorganic Chemistry and Natural Substance, Training and Research Unit of Applied Fundamental Science, Nangui ABROGOUA University of Abidjan, 02 BP 801 Abidjan 02, Côte d'Ivoire

Augustin A. Adima
Laboratory of Water Chemistry and Natural Substance, Training and Research Department of GCAA, Felix HOUPHOUËT-BOIGNY National Polytechnic Institute, BP 1093 Yamoussoukro, Côte d'Ivoire

Henri G. Biego
Laboratory of Biochemistry and Food Science, Training and Research Unit of Biosciences, Felix HOUPHOUËT-BOIGNY University of Abidjan, 22 BP 582 Abidjan 22, Côte d'Ivoire

Anne Marie Niangoran N'guetta
Biochemistry and Food Technology Laboratory, Nangui Abrogoua University, 02 BP 801 Abidjan 02, Côte d'Ivoire

Yolande Dogoré Digbeu
Laboratory of Food Safety, Lorougnon Guede University, BP 150 Daloa, Côte d'Ivoire

Siaka Binaté
Biochemistry and Food Technology Laboratory, Nangui Abrogoua University, 02 BP 801 Abidjan 02, Côte d'Ivoire

Jean Parfait Eugène N'guessan Kouadio
Biochemistry and Food Technology Laboratory, Nangui Abrogoua University, 02 BP 801 Abidjan 02, Côte d'Ivoire

Soumaila Dabonné
Biochemistry and Food Technology Laboratory, Nangui Abrogoua University, 02 BP 801 Abidjan 02, Côte d'Ivoire

Edmond Ahipo Dué
Laboratory of Food Safety, Lorougnon Guede University, BP 150 Daloa, Côte d'Ivoire

A. N. Tsado
Department of Basic and Applied Sciences, Niger State Polythechnic Zungeru, Nigeria

L. Bashir
Department of Biochemistry, Federal University of Technology, P.M.B.65, Minna, Niger State, Nigeria

S. S. Mohammed
Department of Biochemistry, Federal University of Technology, P.M.B.65, Minna, Niger State, Nigeria

I. O. Famous
Department of Biochemistry, Federal University of Technology, P.M.B.65, Minna, Niger State, Nigeria

A. M. Yahaya
Department of Biochemistry, Federal University of Technology, P.M.B.65, Minna, Niger State, Nigeria

M. Shu'aibu
Department of Biochemistry, Federal University of Technology, P.M.B.65, Minna, Niger State, Nigeria

T. Caleb
Department of Biochemistry, Federal University of Technology, P.M.B.65, Minna, Niger State, Nigeria

Volodymyr V. Rushchak
Department of Molecular Oncogenetics, Institute of Molecular Biology and Genetics, NAS of Ukraine,Zabolotnogo Str. 150, Kyiv, Ukraine

Ganna M. Shayakhmetova
Department of General Toxicology, SI "Institute of Pharmacology and Toxicology, NAMS of Ukraine",Eugene Pottier 14, Kyiv, Ukraine

Anatoliy V. Matvienko
Department of General Toxicology, SI "Institute of Pharmacology and Toxicology, NAMS of Ukraine",Eugene Pottier 14, Kyiv, Ukraine

Mykola O. Chashchyn
Department of Molecular Oncogenetics, Institute of Molecular Biology and Genetics, NAS of Ukraine,Zabolotnogo Str. 150, Kyiv, Ukraine

M. A. Al-Mamun
Protein Science Lab, Department of Genetic Engineering and Biotechnology, University of Rajshahi, Rajshahi-6205, Bangladesh

M. A. Hakim
Functional Proteomics and Natural Medicines, Kunming Institute of Zoology, the Chinese Academy of Sciences No. 32 Jiaochang Donglu, Kunming, Yunnan, 650223, China

M. K. Zaman
Protein Science Lab, Department of Genetic Engineering and Biotechnology, University of Rajshahi, Rajshahi-6205, Bangladesh

K. M. F. Hoque
Protein Science Lab, Department of Genetic Engineering and Biotechnology, University of Rajshahi, Rajshahi-6205, Bangladesh

Z. Ferdousi
Protein Science Lab, Department of Genetic Engineering and Biotechnology, University of Rajshahi, Rajshahi-6205, Bangladesh

M. Abu Reza
Functional Proteomics and Natural Medicines, Kunming Institute of Zoology, the Chinese Academy of Sciences No. 32 Jiaochang Donglu, Kunming, Yunnan, 650223, China

Bushra Kalim
Department of Zoology, Government College University, Lahore, Pakistan

Nils Böhringer
Institute for Pharmaceutical Biology, University of Bonn, Bonn, Germany

Nazish Ali
Department of Zoology, Government College University, Lahore, Pakistan

Till F. Schäberle
Institute for Pharmaceutical Biology, University of Bonn, Bonn, Germany

Sonali Hazra Das
Department of Biotechnology, Heritage Institute of Technology, Kolkata, India

Mahasweta Paul
Department of Biotechnology, Heritage Institute of Technology, Kolkata, India

Shankha Ghosh Dastidar
Department of Biotechnology, Heritage Institute of Technology, Kolkata, India

Tulika Das
Department of Biotechnology, Heritage Institute of Technology, Kolkata, India

Srabanti Basu
Department of Biotechnology, Heritage Institute of Technology, Kolkata, India